Gisbert Kapp

Alternate-Current Machinery

Reprinted from the Minutes of Proceedings of the Institution Civil Engineers,

London

Gisbert Kapp

Alternate-Current Machinery

Reprinted from the Minutes of Proceedings of the Institution Civil Engineers, London

ISBN/EAN: 9783337002657

Printed in Europe, USA, Canada, Australia, Japan

Cover: Foto ©berggeist007 / pixelio.de

More available books at **www.hansebooks.com**

ALTERNATE-CURRENT MACHINERY.

BY

GISBERT KAPP, Assoc. M. Inst. C. E.

Reprinted from the Minutes of Proceedings of the Institution of Civil Engineers. London.

NEW YORK:
D. VAN NOSTRAND COMPANY,
23 Murray Street and 27 Warren Street.
1889.

PREFACE.

The subject comprised under the title of this monograph naturally divides itself into the following sub-sections: 1. Alternators; 2. Transformers; 3. Motors; 4. Meters; 5. Mains; and 6. Accessory apparatus for use in central stations and on the premises of the persons supplied with current from such stations. The question of lamps does not properly belong to the subject under consideration, because glow lamps are equally suitable to be fed by alternating and direct currents, and the alterations required to make arc lamps work with alternating currents are easily applied, and present no special interest. In the present volume, the author deals more especially with the first three subjects, namely, alternators, transformers, and motors.

ALTERNATE-CURRENT MACHINERY.

By Gisbert Kapp, Assoc. M. Inst. C. E.

Alternators.

The theory of alternate-current machines, as given in modern text books, is based upon the assumption that the currents are generated in wire coils, the magnetic induction through which undergoes periodical changes according to a simple sine-function. A machine of this character would be represented, in its most elementary form, by a coil of insulated wire revolving round an axis in its own plane, with uniform velocity in a uniform magnetic field, the axis of rotation being at right angles to the lines of the field. Fig. 1 represents such a machine; a rectangular coil A B is placed in the field produced by the polar surfaces N S, and if these surfaces extend beyond the space occupied by the coil, the field within that space may be considered to be uniform. If the axis of rotation passes midway between the sides A and B of the rectangle, the electromotive forces in both branches of the coil are added, and the total electromotive force at any moment is according to a well-known formula—

$$e = 2\pi n z \frac{\tau}{2} \sin a \ldots \ldots (1),$$

where n is the frequency (number of complete periods per second), τ is the number of wires in both branches A and B of the coil collectively, z the total induction through the coil when the latter stands at right angles to the lines of force, and a the angle between this position and the position at which the electromotive force e is generated. The relation between a and t, the time which was required to rotate the coil through this angle, is $a = 2\pi n t$. The total electromotive force changes periodically from a positive maximum through zero to a negative maximum and back again, and the numerical value of these maxima is $2\pi n z \dfrac{\tau}{2}$ —(2). For practical purposes

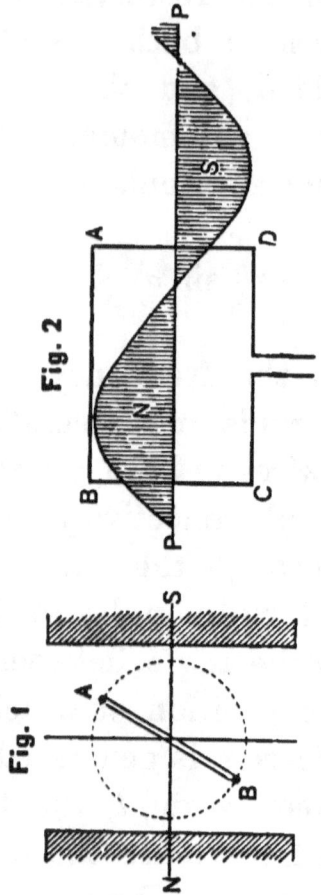

this value is, however, of less importance than the mean value of the electromotive force; for on the latter depends the amount of work obtainable from the apparatus. By mean value of the electromotive force of an alternating current is meant the electromotive force of a direct constant current, which will, in a given resistance free from self-induction (say for instance, the wire of a Cardew voltmeter), produce the same amount of heating. It is necessary that the resistance should be comparatively large, in order to limit the current to such a small value that the disturbing effect of self-induction may be neglected. Let T be the periodic time, then e, the mean value of the electromotive force, is found from the equation—

$$\int_0^{\frac{T}{2}} \frac{1}{R}\left(2\pi n z \frac{\tau}{2} \sin 2\pi n t\right)^2 dt = \frac{T}{2} \frac{1}{R} e^2,$$

when R denotes the resistance of the Cardew voltmeter.

The solution of this equation is a simple mathematical operation, and need, therefore, not be given at length. The result is

$$e = \frac{\pi}{\sqrt{2}} n z \tau \ldots \ldots (3)$$

It is convenient to compare this value with that representing the electromotive force of a direct-current machine having the same field and external number of wires τ (but equally distributed) on the armature. The electromotive force of such a machine is given by the product $z \tau \frac{N}{60}$, where N, the number of revolutions per minute, may obviously be replaced by $60n$ when the machine has two poles, because, in this case, frequency and number of revolutions per second are identical. If the direct-current machine has more than one pair of poles, say, for instance, p pairs, then it is possible to wind the armature in such a way that the electromotive forces due to the different pairs of poles are added, so that the total electromotive force is p times the above value, and the same may obviously be done with the similar machine wound for alternating currents so that in all cases the ratio between the electromotive forces of the two machines remains the same, namely, $\frac{\pi}{\sqrt{2}}$ (4). Of two machines containing the same amount of material, the one which is wound to give alternating currents will therefore produce about two and a quarter times the electromotive force of that wound for continuous current. Since the halves of the armature-conductor in a direct-current machine are grouped in parallel, the strength of current, which such a machine gives when the current density is the same as in the armature of an alternator of equal dimensions, is twice that of the latter, so that the output of the alternator will only be about 12 per cent. greater than that of a direct-current dynamo of equal weight. This result, it should be remembered, has been obtained on the basis of the theory of alternators as found in text-books, whilst neglecting self-inductions, distortions of field, and other disturbing elements; and, before it can be accepted for actual practical work, it is necessary to inquire whether the basis of this theory is in accord with the constructive data to be found in modern machines of this class. Alternators as now constructed are, with a single exception, multipolar machines. This exception is the high-speed dynamo of the Hon. Charles A. Parsons, Assoc. M. Inst. C. E., which can be converted into an alternator by some slight changes in its armature and commutator; but all the other machines now in practical use have more than one pair of poles, which are set in a circle, and are so arranged that the coils of the armature sweep past them. As the interpolar space is always very small, the density of lines, emerging from or entering the polar surfaces, may be taken as fairly uniform over the extent of these surfaces. It thus becomes possible to determine for every shape of coil the configuration of polar surface, which will produce that variation in the induction corresponding to the simple sine-function, taken as the basis of the text-book theory of alternators. In the most simple and most frequently occurring case of a coil with straight sides, it is obvious that the dimension of the pole-piece, which is parallel to the active part of the coil, would have to vary according to a sine-function as shown in Fig. 2, where ABCD represents the coil, and PP the mean pitch line through the poles straightened out. The shaded areas N S are the poles, which are shown on opposite sides of the pitch-line to make their sinoidal contour at once apparent. Poles of this shape are, however, not used in practice; the poles are either trapezoidal, circular, or rectangular. In the Mordey alternator the poles have a trapezoidal face, and those succeeding each other on the same side of the armature are of the same sign, and are placed in line with those of opposite sign on the other side of the armature. Thus along the mean pitch line, uniform fields alternate with spaces in which there is no induction; this arrangement when straightened out may be represented by a succession of rectangles, as in Fig. 3, placed above and below the line PP, the height of the rectangles representing half the induction between the poles. The analogue arrangement of poles in a direct-current machine is shown in Fig. 4, where a cylindrical armature, provided with an iron core, is completely surrounded by its pole-pieces. In the latter class of machines the exact shape of the poles is only in so far of importance as it affects the leakage of magnetism, but for a given total induction through the armature the shape has no influence on the electromotive force. Hence for a direct-current machine, apart from the question of leakage, Fig. 3 is as good an arrangement as Fig. 1, but for an alternator it is not so good. A simple calculation, which need not be reproduced here, shows that the electromotive force of the arrangement represented by Fig. 3 is

$$e = 2 n z \tau \ldots \ldots \ldots (5).$$

In this case the electromotive force is therefore only twice that of the direct-current machine, and as, owing to the concentration of armature windings into one coil, the limit of current as determined by heating will be reached sooner, the current will be somewhat smaller than half that of the direct-current machine, so that the output will be slightly less. It must also be borne in mind that a rectangular field with sharp corners cannot be obtained. In consequence of leakage and spreading of lines, the corners of the field will be more or less rounded, as shown by dotted lines, and this circumstance must lower the electromotive force beyond its theoretical value. On these grounds it would therefore appear that this configuration of field-magnets is not very advantageous, but there are certain other considerations which tend to modify this conclusion. In actual machines it is not a single coil which has to be dealt with, the active wires in which occupy only a very narrow space, but wide coils covering a considerable extent of surface on the armature; there is also the question of self-induction to be taken into account, and the more or less economical way of producing the field. It is especially with

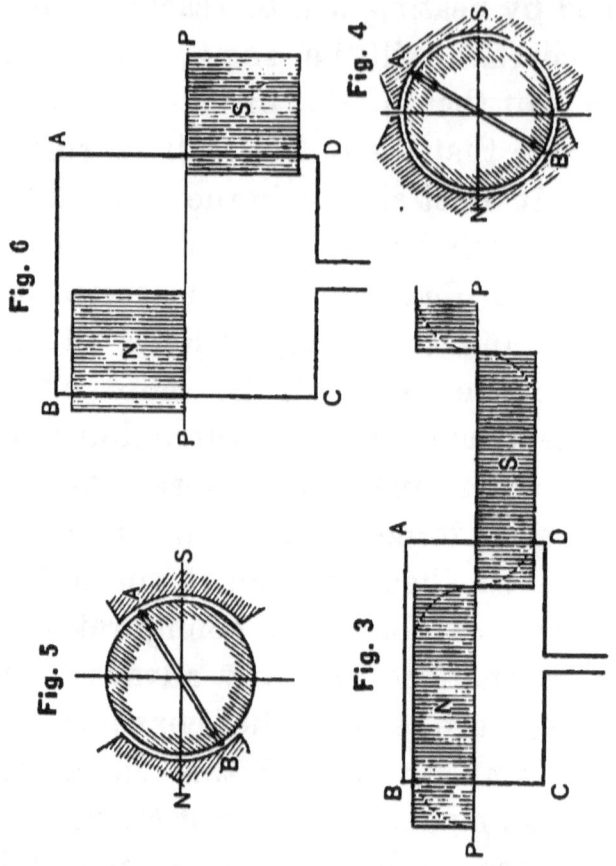

regard to the latter consideration that alternators of this type (of which the Kennedy and Mordey machines are examples) have a considerable advantage, as their fields can be made on the "iron-clad principle," so that no magnetism is lost in external leakage, and only one coil of exciting wire is required.

The type of field most commonly used in modern alternators is, however, one in which N and S poles succeed each other around the armature, and in its simplest form such a machine is represented by Fig. 5, where an iron-cored cylindrical armature revolves between poles which do not completely surround it. The portion of the armature not embraced by the poles may be taken as equal to that embraced, and Fig. 6 shows this disposition in a multipolar machine. PP is the mean pitch-line of the poles straightened out, and the poles are represented by the shaded rectangles which have a width equal to half the pitch. The mean electromotive force of this arrangement is

$$e = \frac{4}{\sqrt{2}} n z \tau \ldots \ldots \ldots (6),$$

or 183 per cent. greater than that of a

direct-current machine containing the same amount of material, whilst the current, owing to the concentration of wires, and consequent greater liability to heat, is something less than one-half. The output of this machine will, therefore, be something less than 40 per cent. greater than that of a direct-current machine having an equal weight. It must not be forgotten that the results here obtained refer to machines in which the active wires on the armature are concentrated into single lines. This is of course impossible in practice, since the wires must be of appreciable thickness, and must therefore occupy a considerable space. An approximation can be made to the theoretical condition by employing an iron-cored armature with projecting teeth, the coils being laid into the recesses. As, however, armatures of this type tend to produce heating of the pole-pieces, if the latter be solid, the arrangement is not generally used; and when the armature is provided with an iron core, the surface of this is in most machines perfectly smooth. The wire coils on the armature, whether this contain iron or not, must therefore occupy a considerable space, and this circumstance will, generally speaking, cause the electromotive force obtainable with the different forms of fields now considered to be lower than the values given. The reason for this reduction is that, with a wide coil not all the turns are equally and simultaneously influenced by the field, and in certain positions the action is differential. The calculation of the effective mean electromotive force for any configuration of field and armature coils presents no special difficulty, being in fact only an extension of the methods already indicated, though for certain complicated forms the operation is laborious. The author only proposes to deal with some simple cases, which may be regarded as forming the limiting conditions of practical designs. As regards the field these cases are:

(1) Rectangular field, width of poles equal to the pitch, as shown in Fig. 4.

(2) Rectangular field, width of poles equal to one-half the pitch, as shown in Fig. 6.

(3) Rectangular field, width of poles equal to one-third of the pitch.

As regards the armature, either the whole available space may be covered with coils, in which case the width of each group of conductors forming a coil, or one side of a coil, will be equal to the pitch, and the armature will contain the greatest possible amount of wire; or, only portions of the available space on the armature may be covered with the active conductors, which plan is generally adopted in modern alternators, because by it the differential action alluded to is diminished. The cases selected for consideration are the following :

(1) Width of coils equal to the pitch.

(2) Width of coils equal to one-half the pitch.

(3) Width of coils equal to one-third of the pitch.

The calculation of the mean electromotive force, for any combination of field and armature of the kind here indicated, is a very simple mathematical operation, and it is not necessary to give it in detail. It will suffice to give the result for those combinations which most nearly resemble the conditions found in actual practice. As has been shown, the electromotive force of an alternator can be expressed as the product of the electromotive force of a direct-current machine, having the same dimensions and weight, and a certain coefficient depending on the configuration of the field-magnets and armature-coils. The winding and other constructive data of an alternator being known, it is only necessary to find this coefficient, in order to determine the electromotive force, by reference to the electromotive force which would be produced by the same machine if the armature-coils were connected in such a way as to give a continuous current. In a paper by the author,* read before this Institution on

---

* Minutes of Proceedings Inst. C. E., vol. lxxxiii. p. 123.

the 24th of November, 1885, it was shown that the electromotive force in volts of an ordinary dynamo is given by the formula:

$$e = z\,\tau\,N\,10^{-6} \quad \dots\dots\dots (7),$$

where z represents the total induction in English lines of force (each equal to 6,000 C. G. S. lines), τ the total number of active wires counted all round the armature, and N the speed in revolutions per minute. If the machine is multipolar, and if the different armature-circuits are arranged in series, the expression for e must be multiplied by p, the number of pairs of poles in the field. Let k be the coefficient which expresses the ratio between the electromotive forces of the alternate and continuous-current machines, then the formula for the former is—

$$e = k\,p\,z\,\tau\,N\,10^{-6} \quad \dots\dots (8)$$

if the total induction be given in English lines of force. If in C. G. S. units the electromotive force is—

$$e = k\,p\,z\,\tau\,\frac{N}{60}\,10^{-8}.$$

The following table gives the value for k for different cases, all referring to poles of rectangular shape and coils in which the active wires are straight:—

1. Width of poles equal to pitch, toothed armature and winding concentrated in the recesses . . $\}$ $k = 2.000$
2. Width of poles equal to pitch, smooth armature and winding spread over the whole surface . $\}$ $k = 1.160$
3. Width of poles equal to pitch, smooth armature and winding covering only one-half the surface . . . . $\}$ $k = 1.035$
4. Width of poles equal to half the pitch, smooth armature and winding spread over the whole surface . . . . $\}$ $k = 1.635$
5. Width of poles equal to half the pitch, smooth armature and winding covering only one-half the surface . . . . $\}$ $k = 2.300$
6. Width of poles equal to one-third the pitch, smooth armature and winding covering only one-third of the surface . . . . . . $\}$ $k = 2.830$

According to the ordinary sine-formula the coefficient is $k = 2.220$; and this agrees fairly well with case 5, which is the most frequently met with in actual practice. The formula for e presupposes the knowledge of the total induction z, which depends upon the shape and size of the field magnets, their arrangement, the amount of iron in the armature, the interpolar space, and the exciting power. The time has long passed when a consideration of the relation between exciting power and total induction would have been of interest. This subject has been fully treated by Drs. J. and E. Hopkinson before the Royal Society,* and the author has also dealt with it in a paper on "The Predetermination of the Characteristics of Dynamos," before the Society of Telegraph-Engineers and Electricians.† These methods have subsequently been improved, by the introduction of certain terms due to Professor Forbes, by means of which the magnetic leakage can be more accurately determined. With the knowledge of the total induction z, and the coefficient k, it is thus possible to determine the relation

---

* "Dynamo-Electric Machinery," by Drs. J. and E. Hopkinson Phil. Trans. Royal Society, 1886, p. 331.
† Journal, vol xv. 1886, p. 518.

between exciting-power and electromotive force of any given alternator, provided the current flowing through its armature is not so great as to produce a sensible self-inductive effect; in other words, the formula renders it possible to determine the static, but not the dynamic, characteristic. The effect of self-induction is to produce an electromotive force, the phase of which is at right angles to that of the current; and the electromotive force available for doing work in the circuit is the resultant of the induced electromotive force and that due to self-induction. The mathematical treatment of the problem is of great difficulty, not only because, with the shape of poles and armature coils occurring in modern alternators, the electromotive force is a very complicated function of the time, but also because the coefficient of self-induction is not a constant, but varies with the relative position of the coils and poles. Fortunately the coefficient is comparatively small. In machines containing no iron in the armature, such as the Mordey and Ferranti, it is almost negligible; in machines containing iron in the armature, which is not encircled by the coils, such as the Westinghouse and Lowrie-Parker, it is still very small; and even in machines in which the coils are wound round the armature-core, as in Kennedy's machine and that designed by the author, the coefficient, although appreciable, is yet not so large as to cause a sensible error being introduced by the assumption that it is a constant.

The method originally devised by Mr. Joubert, for taking the self-induction of an alternator into account, requires the solution of a differential equation, but the problem can be treated graphically in a much more simple manner. Both methods have this in common, that it is assumed that the current and all electromotive forces are sine-functions of the time. Strictly speaking, this is not correct, but the error is probably not very great, and it is in a certain sense unavoidable; because if this assumption were discarded, and each case treated on the basis of the exact induction-curve as mapped out from the shape of the poles and armature-coils, it would lead to such difficulties as to make the investigation useless for practical purposes, for which easily understood and simple methods of calculation or construction are wanted. The sine-function has, therefore, as far as the author is aware, always been taken by previous investigators as the basis of their calculations, and in this he proposes to follow their example.

Let, in Fig. 7, the length of the line O I represent the maximum value of the current (crest of the current wave), and let this line revolve in the direction of the arrow around O as the center, with an angular velocity $w = 2\pi n$, then the projection of O I upon the vertical gives the current at any moment. Let the line O L represent the maximum electromotive force of self-induction due to the current O I, then the projection of O L will similarly give the electromotive force of self-induction at any moment, the two lines preserving during rotation their rectangular position. The effective electromotive force which produces the current O I, through a resistance having no self-induction, may be represented by a length O E drawn to the same scale as O L, and since the effective electromotive force must be the resultant of the induced electromotive force and the

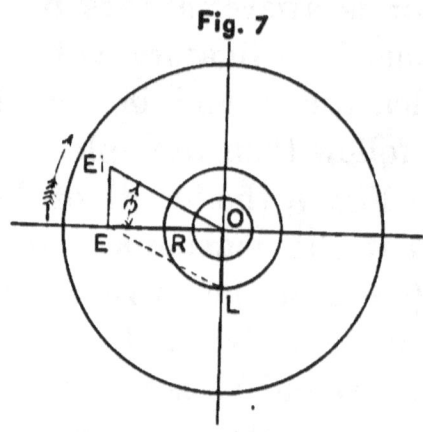

Fig. 7

electromotive force of self-induction, the former can be found by constructing the parallelogram O L E E_i, the line O E_i giving the induced electromotive force both as regards magnitude and position. If the length O R represents the electromotive force lost on account of the resistance of the armature, the remaining portion R E represents the electromotive force actually available for doing work through the resistance of the external circuit; and this will the more nearly approach the induced electromotive force the smaller the armature resistance, and the smaller the self-induction of the whole circuit. In order to obtain a maximum of work in the external circuit with an alternate-current plant, it is, therefore, advantageous that the armature of the alternator should have a minimum of resistance, and that the total self-induction of the circuit should be small. The commercial value of a given type of alternator depends, amongst other things, upon the amount of work which can be obtained from a given weight of materials employed in the construction of that particular type of alternator. This value will, therefore, be roughly indicated by the ratio between the length of the lines R E and O E_i; and this ratio the author proposes to call "plant-efficiency." It must be borne in mind that the plant-efficiency is not an indication of the more or less perfect way in which the alternator transforms mechanical into electrical energy, but merely an indication of the commercial value of any given type for producing a maximum amount of external work with a minimum weight and cost of the materials employed. The electrical efficiency of the machine can also be found from the diagram. According to a well-known law, the internal energy of the machine is equal to half the product of maximum current and maximum induced electromotive force multiplied with the cosine of the angle of lag Φ between the two.

Internal energy $= \frac{1}{2} E_i I \cos \Phi$,
when E_i is the maximum value of the induced electromotive force and I the maximum value of the current. The external energy available for doing work is the product of the mean current and mean terminal electromotive force, which is equal to half the product of their maximum values.

External energy $= \frac{1}{2} E I$,
when E is the maximum value of the

terminal electromotive force. The electrical efficiency is therefore given by the ratio of the lines R E and O E, which is obvious considering the electromotive force O R, absorbed in overcoming the resistance of the armature, is the only purely electrical loss occurring in the armature.

In alternators used for parallel distribution, the object aimed at is generally to keep the terminal pressure constant for all currents, and this can be attained by working with a constant field and suitably varying the speed, or by maintaining a constant speed and adjusting the exciting current so as to suitably vary the electromotive force. The latter plan must be adopted if several machines are worked in parallel; and in this case the diagram, Fig. 7, may be used to determine the limits between which the exciting current must be varied, to keep the terminal electromotive force constant for all loads from no current to the greatest current the armature is intended to pass. The mean induced electromo-.

tive force e is found by Formula 1 and its maximum value is obviouly—.

$$E_i = \sqrt{2}\, e \quad \ldots \ldots \ldots \quad (9).$$

The maximum value of the electromotive force of self-induction is given by—

$$E_s = 2\pi n \text{L I} \quad \ldots \ldots \ldots \quad (10),$$

where L is the coefficient of self-induction, which can be determined by various well-known methods; or by running the armature in a constant field, and measuring the terminal electromotive force with and without a current passing through the armature. The electromotive force lost in resistance is simply the product of the current I and the resistance R of the armature. With the knowledge of these quantities it is now easy to determine the strength of field required for any current, since, according to Formula 1, the field is proportional to the induced electromotive force. For this purpose the diagram, Fig. 7, may be used, or the modified form shown in Fig. 8. Here O A represents the terminal electromotive force, which is to be kept constant. A

B represents to the same scale the loss of electromotive force in the armature when the machine is giving its full current, and O D the corresponding electromotive force of self-induction. Make B C parallel with and equal to O D, and describe circles over the diameters O A, O B, O D, and O C; then a straight line

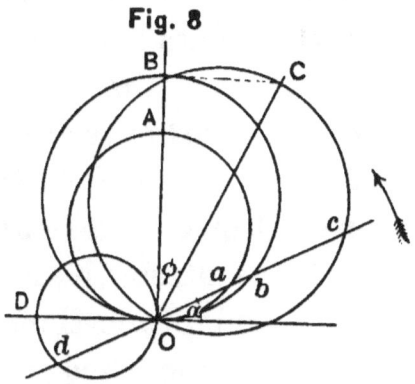

Fig. 8

revolving round O as a center, in the direction of the arrow, is cut by these various circles in such a way that the segments represent the respective values of the different electromotive forces at any moment. Thus, at the time when the angular position *a* has been reached, the induced electromotive force is O*c*, the effective electromotive force is O*b* and the terminal electromotive force is Oa, all three being positive, that is, in the direction of the current, whilst the electromotive force of self-induction is negative, and is given by the length Od. As the load decreases, the circle O D becomes smaller, until, when no current is allowed to flow, it has shurnk into a point, whilst there is no loss due to resistance, and the points B and C coincide with A. In this case the terminal and induced electromotive forces are identical, and the strength of the field is a minimum and proportional to the length O A. At full output the strength of the field is a maximum and proportional to the length O C. Having thus found the limits of the strength of field required in these two cases, it is an easy matter to determine, by the usual formulas connecting strength of field and exciting power, the strength of the exciting current for each. It will be immediately apparent from the diagram, that of two machines, the one having a high and the other a low plant efficiency, the latter will at full output require more exciting current, and if the resistance of its field be the same, this circumstance will influence the electrical efficiency unfavorably. On the other hand, a sensible amount of self-induction, although it does reduce the plant-efficiency, is indispensable for the safe working of alternators in parallel, and must, therefore, be regarded as rather an advantage than otherwise.

The problem of parallel working is of the greatest importance for central stations, as only by such an arrangement of machines can absolute continuity of the service and the greatest economy at all times be ensured. Alternate-current distribution should, if possible, be carried on in the same simple manner as direct-current distribution, that is to say, the number of machines at work should correspond as nearly as may be to the output at any time; the addition or withdrawal of machines should not even momentarily interrupt the supply of current to any part of the system of distribution. These conditions, it will easily be seen, can only be fulfilled if the machines are capable of working in parallel, and it is, therefore, of considerable practical importance to investigate the conditions under which alternators may be expected to work safely in parallel. This subject is closely allied with that of working alternators as motors, and it is, therefore, convenient to consider the two problems jointly. The analytical treatment of an alternate-current motor is even more difficult than that of an alternate-current generator; but by an extension of the graphic method, the question can be treated in a very simple and easily understood manner. It is well known that an alternator will not start without mechanical assistance, but that, having been started at a speed approximately corresponding to the frequency of the supply current, it is kept in motion when this current is switched on to its terminals, and it may under certain conditions even give off mechanical energy. The problem involved may therefore be stated somewhat in this fashion: Given an alternator, with excited field and running at a proper speed, and a pair of terminals from which any desired strength of current can be obtained at a constant pressure, and having a frequency approximately corresponding to the speed of the alternator, what will be the condition under which the alternator will fall into step with the supply-current, and what will be the relation between the current passing and the mechanical energy given off? Also, how will this relation be affected by variations in the strength of the field? Let, in Fig. 9, the circle E_t represent the terminal electromotive force, and the circle E the electromotive force induced in the armature; and assume a certain current under which the electromotive force of self-induction is O L and that lost in resistance O R. Since both quantities are proportional to the current, their resultant O A will for all currents form the same angle with the axes, the point A being simply shifted further out on the line Oa if the current increases. It is obvious that the line O A can also be considered as the resultant of E and E_t, and the position of these two electromotive forces becomes thus at once defined. In order that the machine may give off mechanical work, the induced electromotive force must

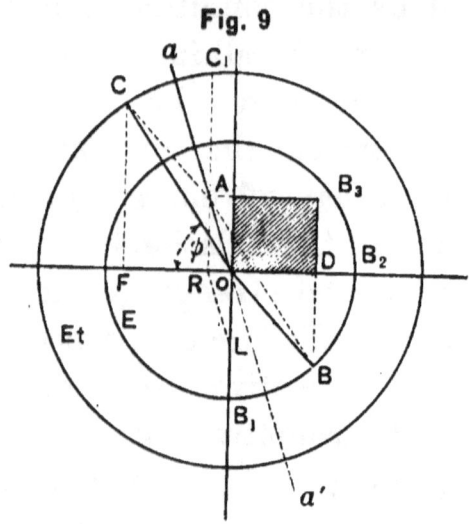

Fig. 9

evidently be opposed to the current, that is to say, it must lie to the right of the vertical, whilst the terminal electromotive force must lie to the left. Under these conditions only one parallelogram of forces is possible, namely, that determined by the points O B A C where

$OB = E$ and $AB = E_t$. The mechanical energy given off is proportional to the product O A and O D, the latter quantity being the horizontal projection of O B, and obviously equal to R F. Now consider what will happen if some of the load is taken off the machine. The immediate result will evidently be a slight acceleration of the armature, by virtue of which the maximum electromotive force E is induced sooner than previously. In the diagram this effect is represented by the advance of the point B towards B_1 and a corresponding displacement of C towards C_1. The length O A will also be slightly altered, viz., decreased until B has come into line with line Oa', and further on increased again. If the whole of the load be taken off, B will almost coincide with B_1 and the angle of lag Φ will be a maximum; but the length O A, that is to say, the current, will not be materially different from its previous value, as can easily be seen from the diagram. Hence it follows that, even when running light, the motor will allow a considerable current to pass; but, as the phase of the current lags behind that of the terminal electromotive force by nearly a quarter period, the energy supplied is very small; in fact, only sufficient to overcome the resistance of the armature-circuit and mechanical friction. With an increase of load the armature will be slightly retarded, which will bring the point B further away from B_1 in the diagram. Simultaneously the lag Φ will decrease, and the current which is proportional to the length O A will increase, the increase being slow until the point B_2 is reached, and afterwards more and more rapid. Since the mechanical energy given off is proportional to the product of current, and induced electromotive force multiplied by the cosine of the angle the latter forms with the current, it may be represented by the area of the shaded rectangle, and a series of such rectangles for different positions of the point B on the circle, which represents the induced electromotive force, can be determined.

In this way it is found that the mechanical energy continues to grow as B is shifted higher up on its circle, until a point in the neighborhood of B_2 is reached, after passing which the energy decreases again. Any position of B between B_1 and B_2 is stable, an increase or decrease of load resulting in an automatic adjustment of the current and of the phases of the electromotive forces; but if the machine be overloaded beyond the point B_2, its condition is unstable, as an increase of load, tending to further retard the induced electromotive force, must necessarily reduce the capacity of the machine for doing work, and thus bring it out of step with the current. Here is a danger from which direct-current machines are exempt. The power of a direct-current machine, when used as a motor, increases indefinitely with the current, and an overload, if it does not last long enough to burn up the armature, does no harm; but with an alternator, an overload, if it lasts only a few seconds, may bring the armature to a stand-still, and if the self-induction is not sufficient, this may eventually cause it to burn up.

It has been shown by Mr. H. Wilde* and by Dr. J. Hopkinson,† that two similar alternators can be coupled in parallel to the same circuit; and Professor Adams described, in a paper read in 1884 before the Society of Telegraph-Engineers and Electricians,‡ some experiments in which he succeeded in driving an alternator as a motor, current being supplied by another alternator of similar construction. Mr. Westinghouse, in some central stations in America, has also worked alternators in parallel. There is thus sufficient experimental proof that such a method of working is possible, and the only question which remains to be investigated is whether, for safe working, any very

---

* Proceedings Lit. and Philosophical Society of Manchester. Vol. viii. p. 62; and "Philosophical Magazine," January, 1869. Fourth Series. Vol. xxxvii. p. 54.

† The Inst. C. E. Lectures on "The Practical Applications of Electricity." Session 1882-83. "Some Points in Electric Lighting." By Dr. J. Hopkinson.

‡ Journal, vol. xlii. p. 515.

delicate adjustment of the two machines is necessary. If such were to be the case, it is obvious that the system would not be suitable for use in central stations, where the service must be carried on day and night, and perferably, by a staff not composed of skilled electricians. If a battery of alternators be already running in parallel, and it be desired to add another machine, the obvious precaution which an ordinary attendant would take is to excite its field, and to run the machine up to the proper speed before joining it to the circuit. Some devices have been introduced for indicating the coincidence of phases by means of glow-lamps, but as the use of these requires careful attention and quickness of perception on the part of the operator, it would be preferable to so construct the machines that absolute coincidence of phase should not be required. If each alternator is driven by its own engine, and if the latter is provided with a good governor, there will be no difficulty in obtaining very approximately the right speed, and the only adjustment which must be left to the attendant is that of the exciting current. The question now arises, within what limits an error in this adjustment may be safe. To this question the diagram, Fig. 10, gives the answer.

In this diagram the circle E_t represents as before the terminal electromotive force, and the line Oa the direction of the resultant of the electromotive force of self-induction, and that lost in overcoming the resistance of the armature. The coefficient of self-induction depends upon the density of lines within the armature core, being obviously the smaller, the more nearly the core is magnetized to saturation. For a reason which will be explained presently, the iron in the armature of alternators is never magnetized to any considerable extent, the induction seldom exceeding about 7,000 C. G. S. lines, and as at this low figure the permeability is not much, if at all, affected, it can be assumed, without any great error, that the self-induction is a constant for all field intensities which occur in practice. The angle which the line Oa forms with the horizontal may therefore be taken as unaffected by the exciting current, and the length O A as proportional to the current passing through the armature.

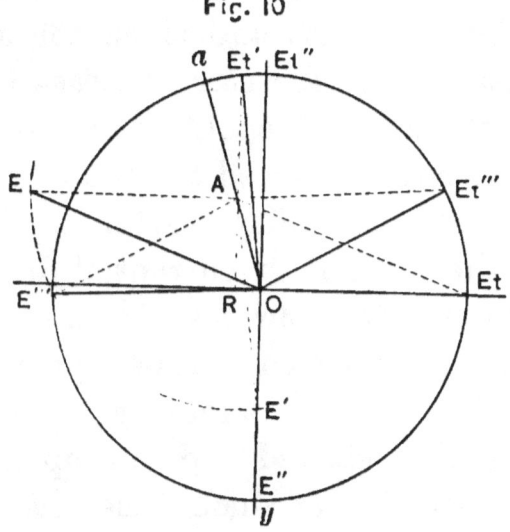

Fig. 10

In fact, by employing a suitable scale the armature current may be read directly off this line on the diagram. Suppose then that O A represents the greatest current which may be safely allowed to flow. The determination of this current will generally not depend so much on the capacity of the machine which is being switched on, as on the consideration as to how much current may be temporarily taken from the other machines at work. A new machine is only added, if the limit of output of these machines is in danger of being overtaken by the demand for current in the whole system; and since the available margin will therefore necessarily be small, the machine about to be coupled up must not be allowed to act as a heavy drain upon the other machines. This limits the amount of current at disposal for bringing the new machine into step. Let this maximum current be represented by the line O A. Before switching on, the engine is started and run up to the proper speed, in which case the only work done by the engine is that of overcoming its own friction, and the frictional resistance of the machine. If, at the moment of switching on, the induced electromotive force occupies a position to the left of the line Oy, there will be a momentary increase of current, and a considerable retarding action of the armature, causing it to lag so that the induced electromotive force will immediately be shifted back to a position on the right of Oy, and, finally, as the engine is supposed to give all the power required for running light, equilibrium will be attained by the induced electromotive force occupying a vertical position, when no work is being done by the current. The question now is, what must be the induced electromotive force to limit the current to the value of OA? This is easily found by describing round A as center with radius E_t, a circle, and marking its point of intersection E' with the vertical Oy. That strength of field which will give the induced electromotive force OE' is the least which may safely be permitted. The terminal electromotive force will then occupy the position OE'_t, which line is parallel with and equal to E'A. It should be observed that there is between the terminal electromotive force and the current a difference of phase, amounting to nearly a quarter period. This circumstance tends greatly to minimize the disturbing influence of the machine, which absorbs current upon the other machines supplying current to the mains. Thus the idle machine takes most current at a time when the mains take hardly any, and when the mains take maximum current that passing through the machine is nearly zero. This automatic process of compensation is valuable, as it allows the permissible current, for bringing the new machine into step, to be fixed considerably higher than would otherwise be safe. If, without increasing the field, more steam be given to the engine, causing it to push the armature into a more advanced phase, the point E' will move to the left of the vertical, and E'_t will move to the right, whilst the current will increase. As soon as E'_t has passed the vertical it opposes the current, and the machine ceases to act as a motor, but acts as a generator supplying current to the mains. It is thus possible to run the machine as a generator in parallel with the other machines, notwithstanding the fact that its field is considerably weaker; but it will readily be understood that the working under these conditions cannot be economical, though it may be perfectly safe. Now suppose that, instead of giving more steam to the engine, the strength of the field is increased. The point E' will then be shifted further down on the vertical line Oy, and the point A will be similarly shifted further towards O, that is to say, the current will be decreased. When the field has been so far strengthened as to make the induced electromotive force equal to the terminal electromotive force, the point E' will occupy the position E'', and the terminal electromotive force the position OE''_t, whilst no current will be passing. If now more steam is given, the machine will begin to work as a generator, and the current will reappear and attain its former value, when the power given off by the engine corresponds to the position E''' of the induced electromotive force. The terminal electromotive force will then occupy the position E_t'''. Under these conditions the machine will work as a generator, and much more advantageously than formerly, though not yet with the best plant efficiency. To obtain the best result the field must further be strengthened, until the induced electromotive force occupies the position E, and the terminal electromotive force the position E_t, when the maximum output is reached, and the machine is in exactly the same condition as the other machines. The safe limits, within which the strength of the field may be varied, are given by the length of the lines OE and OE', and the ratio of the corresponding exciting currents must obviously be still greater than that between these two lines. It is instructive to observe what part armature resistance and self-induction take in determining this ratio. As regards the armature resistance, the lower this can be made the better; in modern machines the resistance is generally so small that the loss of pressure is limited to from 2 to 5 per cent. of the terminal electromotive force. The length of the line OR in the diagram will therefore be, at most, one-twentieth part of OE. When the self-induction is negligible, the point A will almost concide with O, and E′ will almost coincide with E″, which shows that the safe margin between minimum and maximum field-strength, and, consequently, also between the smallest and largest permissible exciting current, is extremely narrow. Machines of this type can only be run in parallel if the strength of their fields is adjusted with almost mathematical precision, and as this would require more skill and attention than is available with the ordinary staff of a central station, such machines are practically unfit for parallel working. To make them fit for this method of working either the armature resistance or the self-induction must be increased. An increase of resistance, in order to be effective, would have to be so considerable (as is easily seen from the diagram), as to seriously prejudice the electrical efficiency of the machine, and this expedient may, therefore, be dismissed as impracticable. The other plan of increasing the self-induction is not open to the same objection. It has the effect of lowering the plant efficiency, but its influence upon the electrical efficiency is only indirect, and so small that it may be neglected. From the foregoing, it will be readily seen that the only and sufficient condition, for successful parallel working, is a sensible amount of self-induction in the armature circuit. If the armature itself does not possess this quality in a sufficient degree, a choking coil of suitable self-induction must be inserted into the circuit of each machine. The results here arrived at, by a mere theoretical investigation, are entirely borne out in practice. It is well known that alternators having no iron in their armatures cannot be run in parallel, except by the adoption of some such expedient as choking coils; also that parallel running is feasible with those alternators which have iron-cored armatures, and then with different degrees of security. To make this point clear, compare two machines of different types such, for instance, as the Westinghouse or the Lowrie-Parker on the one hand, and the Kennedy or the author's machine on the other hand. In the two first-named machines the coils are simply laid on to the surface of the armature-core on one side, whilst in the other two machines they completely surround the core. The coefficient of self-induction may, in the latter case, be roughly taken as four times that of the former case, other conditions being approximately the same; and it may therefore be expected that parallel working will be safer and more certain when the armature conductor surrounds the core, than when it lies only on one side of the core, though in this case it will still be possible. The experience gained in America with Westinghouse machines is in this respect very instructive. It has been found that the experiment of parallel coupling always succeeds when the machines are loaded to about half their maximum output or over, but that machines working under a less load cannot, with certainty, be so coupled. A glance at the diagram, Fig. 10, shows the reason for this. The margin of field strength is represented by the difference in the length of the lines OE′ and OE. This difference depends, to a great extent, upon the length of the line OA, which is proportional to the current. If the current through the machines already at work is small, it will be further reduced as soon as the new machine is switched on, and under these conditions the margin of field strength may become so small, that a slight reduction of the exciting current of one machine may cause this to be overpowered by the other machines, and hence the experiment may fail. If, however, the current through all the machines is sufficiently great, then the margin will also be sufficient to cover such irregularities in the exciting currents, and in the torque of the engines, as may be expected to exist when the plant at a station is in the charge of ordinary attendants, who are not skilled electricians or engineers, and hence the experiment will succeed.

It has now been shown how, by the aid of certain diagrams, the behavior of any type of alternator, as regards its ability to run in parallel, can be ascertained; but, as the use of these diagrams involves a certain amount of study, it would be desirable if they could be superseded—at least for ordinary practical work, where an approximate solution of such problems is quite sufficient —by some more simple method. Such a method is afforded by the use of the characteristics of alternators. If a machine of that class be run at constant speed on open circuit, and with different strengths of exciting current, different terminal electromotive forces are obtained. A characteristic can then be plotted, the abscissas of which represent exciting power, whilst the ordinates represent terminal electromotive force, which in this case is the same as the induced electromotive force. Having obtained this curve, which the author suggests should be called the static characteristic, the dynamic characteristic of the machine is determined for a certain current through the armature, by so adjusting the resistance in the external circuit, that for all exciting powers the current remains the same. It is important that the external circuit should not contain any appreciable amount of self-induction—a condition easily fulfilled if, during the experiment, the machine be set to work a bank of transformers which are feeding glow-lamps. The curve so obtained must obviously lie wholly under the first curve, and the difference between the ordinates of the two curves depends upon the choice of current for which the second test has been made, and upon armature resistance and self-induction. Besides these two curves there is a third, namely, that obtained when the machine is driven as a motor running light. To find this curve it would be necessary to have at disposal at least two similar machines, using the one as generator and the other as motor, and during the test the field strength of both machines would have to be regulated, so as to keep the current at its predetermined value. Thus, a "motor characteristic" would be obtained, lying wholly above the "static characteristic." As the last experiment is, however, only possible if two machines can be tested at the same time, and is also somewhat delicate, another method of investigation is desirable. This can be done by using diagram Fig. 10, to deduct the motor characteristic from the two other curves, the experimental determination of which presents no difficulties whatever, and would, indeed, be made as a matter of course in the ordinary test of the machine before it is sent out from the manufacturer. The static characteristic gives the strength of field for every exciting power. From the measured resistance of the armature it can easily be found how much of the effective electromotive force has been absorbed in overcoming this resistance, whilst the dynamic characteristic shows the terminal electromotive force. Thus all the necessary elements exist for determining, by the aid of a diagram similar to Fig. 10, the length of the line O A and its angular position. Having found the point A, the position of the point E' is determined, that is to say, the strength of field O E', which in a motor will allow the predetermined current to pass. The corresponding exciting power is found from the static characteristic, and this gives one point of the motor characteristic. In the same manner other points corresponding to different terminal electromotive forces can be ascertained. The construction is a mere geometrical operation, and need, therefore, not be explained in detail. Fig. 11 shows the relative position of these curves for one of the author's machines, constructed for an output of 60 kilowatts. For the figures from which the diagram has been constructed the author is indebted to Mr. C. E. L. Brown, who has kindly consented to carry out an independent series of experiments, at the works of the Maschinenfabrik Oerlikon, in Switzerland, the electrical department of which is under his charge. The experiments were made as follows: The alternator was driven at its normal speed of 600 revolutions per minute by a steam engine, and its terminals were connected with a

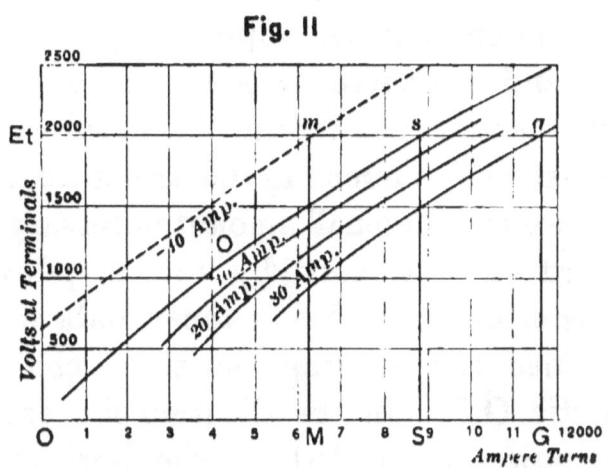

Fig. II

transformer wound to reduce the pressure in the ratio of 20 to 1. The pressure on the secondary terminals of the transformer was measured by one of Sir William Thomson's balances, specially made for alternate-current work. The alternator was connected with a bank of transformers, and the secondaries of the latter with lamps and other resistances. A Siemens dynamometer was inserted into the armature circuit, and the secondary currents from the transformers were occasionally measured by a Sir William Thomson's ampere balance. The primary and secondary currents were found to be very nearly in the ratio of 1 to 20. The exciting current was adjusted by a rheostat, and measured by a spring-solenoid instrument. The first run was made without external current in the armature circuit, except the small amount necessary to record the pressure by the stepdown transformer. In this manner, data were obtained for plotting the static characteristic O S. In the diagram the exciting power is plotted on the horizontal and the pressure on the vertical. The exciting power was found by multiplying the exciting current by 280, there being one hundred and forty turns of wire on each magnet-coil. The resistance of the armature when warm is 1.74, and that of the field 1.73 ohm. The bank of transformers, with its load of resistances, was next switched on, and the latter so regulated as to give an armature current of 10 amperes at different degrees of excitation. The dynamic characteristic marked " 10 amperes " in the diagram was thus found. Similar experiments were made to obtain the characteristics at 20 and 30 amperes, which are also shown. Either of the three dynamic characteristics can now be used to determine, by the aid of diagram Fig. 7, the coefficient of self-induction of the armature; and, if it were not for certain disturbing causes, the three coefficients thus obtained should be identical. In reality, this is, however, not the case, chiefly because the impressed electromotive force is not a true sine function of the time. There is also another disturbing element, namely, the demagnetizing influence of the armature current upon the field. When the machine is working as a generator, the poles developed in the armature by the current itself are of the same sign as the field-poles which they approach, and the armature has therefore a tendency to push back some of the lines emanating from the field-poles. To overcome this tendency the exciting current must be increased, the effect being similar to, though not identical with, that produced by self-induction. When the machine is working as a motor, the effect of the armature current is generally, but not always, to strengthen the field-poles. The author has selected the dynamic characteristic at 30 amperes, for the determination of the coefficient of self-induction; because the great distance between this curve and the static characteristic made it probable that the result would be more accurate than with one of the nearer curves. The coefficient of self-induction was found to be $L = 0.955$ in C. G. S. measure, and 0.0955 in practical units. Taking this as a basis, it is now an easy matter to find, by means of the graphic construction shown in Fig. 10, the motor characteristic for any given current. A current of 10 amperes was chosen, as sufficiently small to avoid a sensible disturbance of the working machines when a new machine is switched on, and this gives the dotted curve, marked "— 10 amperes," in Fig. 11. The interpretation of this diagram is obvious. The normal terminal pressure of this machine is 2,000 volts, and by drawing the corresponding horizontal line, its points of intersection with the curves $m \ s \ g$ fix the exciting power in each case. In order that the machine may work as a generator, giving its maximum current of 30 amperes, the exciting power must be O G. If the machine is merely to run without producing or absorbing current, it must be excited to an amount represented by O S, and if running idle as a motor, and allowing not more than a 10 ampere current to pass, the excitation must be O M. The ratio between the length of the lines O M and O G gives, therefore, in a rough-and-ready way, an indication of the degree of safety when working in parallel.

The formulas and diagrams given above

will be found sufficient for the purpose of designing alternators, or for the determination of the behavior of any given machine of this class, except in one particular, viz., its greater or lesser liability to heating. In any dynamo machine there are two causes for heating, in addition to purely mechanical friction. These are the resistance of the conductors, and the change of magnetization of the iron parts of the machine. As regards the first cause, the result of which may be comprised under the term "copper heat," it is so well understood that it need not in this place be specially considered. One circumstance in connection with this subject should, however, be mentioned. The liability to the generations of eddy currents in the armature conductor is much greater with alternators than with continuous-current dynamos, and in alternators it varies with the construction of the armature to a greater degree. When the armature contains no iron, the heating from eddy currents is much greater than when such a core is used, and the armature conductor must therefore be made in the form of a very narrow strip; so narrow, in some cases, that two or more conductors must be placed in parallel to obtain the required cross-sectional area. The author is unable to suggest an entirely satisfactory explanation for the effect of the iron core in reducing eddy currents in the copper, and merely mentions the fact as deserving attention. In direct current machines he has observed a somewhat similar phenomenon. When the pole-pieces are provided with extensions, the heating of the armature bars is much less than when these extensions are absent, and consequently the trouble of heating can to a certain degree be overcome by the simple expedient of fitting so-called "horns" to the pole-pieces. Another cause of the generation of heat, in machines of all types, is the change of magnetization in the iron parts. This can be minimized by lamination, but never entirely avoided. In direct-current machines the heating from this cause is, however, generally so small that no account is taken of it, except when the induction in the armature core is pushed beyond 18,000 or 20,000. In alternators a much lower induction already produces sensible heating, and it is for this reason that the maximum induction must be fixed at a very low figure. With a perfectly laminated core, the heating is attributed to an effect which Professor J. A. Ewing calls "hysteresis," and which may be described as molecular friction. This scientist has experimentally determined the energy required to carry certain samples of iron through a complete cycle of magnetization of varying degree, and the results have been published in the Philosophical Transactions of the Royal Society.* They are given in ergs per cubic centimeter; but as these units are rather inconvenient for application in the workshop or drawing office, the author has, in the following table, which refers to annealed wrought-iron, translated them into watts per ton.

---

* Phil. Trans. Royal Society, 1885, p. 523.

Induction.	Watts per ton $n = 100$.	HP. wasted in Heat.
2,000	650	0.87
3,000	1,100	1.48
4,000	1,650	2.21
5,000	2,250	3.02
6,000	2,900	3.89
7,000	3,750	5.03
8,000	4,450	5.97
9,000	5,550	7.43
10,000	6,650	8.90

A third column has been added, showing the HP. transformed into heat, and therefore wasted per ton of armature-core, if the frequency is 100 complete periods per second.

In the paper referred to above, Professor Ewing says that mechanical vibrations tend to decrease the amount of energy dissipated per cycle; and that, as the armatures of dynamo machines are, whilst working, in a continual state of vibration, it will be very much less than indicated by the figures he found when experimenting on pieces of iron which were at rest. The author, however, doubts whether this view can safely be accepted. The vibrations of an armature are not of the sharp character comparable to a series of blows, which would tend to shake the magnetism out of the core, and practical experience shows that with high inductions there is a very considerable amount of heating. In an experimental machine which the author constructed, the induction through the armature core was about 16,000, and with a frequency of 80 this core became heated far beyond the point which might be considered safe in practical work. In subsequent machines the induction was therefore reduced by degrees, until a safe limit was reached at about 7,000 lines per square centimeter.

Professor Ewing found that the energy dissipated by hysteresis increased with the speed at which the cyclic change of induction was performed. This circumstance has not been taken into account in the preparation of the table, as no exact data are at present available, by which the increase of heat due to what Professor Ewing calls "viscous hysteresis" can be determined. There is, however, good reason to believe that the speed, at which the cyclic change of magnetism is performed, has a considerable influence on the amount of energy dissipated in heat per cycle. It is probably due to the disadvantage of too great a speed that most makers of alternators find it advisable to work at a moderate frequency. Another reason telling against high frequency is that either the speed of the machine, or the number of field-poles, or both, must be inconveniently great. A very high speed is mechanically objectionable, and the adoption of a large number of field-poles has the disadvantage of reducing the pitch, and therefore of increasing the magnetic leakage, whilst at the same time lowering the coefficient k, all of which tend to an increase in the weight, bulk, and cost of the machine. On the other hand, too low a frequency would also tend to increase the cost of the machine, and there must therefore be, for each type of alternator, a particular frequency at which the best results are obtained. Existing knowledge of the subject is not sufficient for the determination of this frequency by a mere theoretical investigation, but the experience obtained by several makers of these machines, during the last few years, tends to show that the best frequency for nearly all types is below 100. Something more will be said on this point later on.

Transformers.

The theory of transformers is so simple, and so generally understood, that it would be waste of time to enter into it at length in this place. Modern transformers are provided with a core of laminated iron, which is symmetrically placed in respect to the two windings, so that the same induction passes through both the primary and the secondary coils. Under these conditions, the electromotive force induced in the two circuits is simply proportional to the number of convolutions they respectively contain, and the terminal electromotive forces are found by adding in the case of the primary, and subtracting in the case of the secondary, the loss of pressure due to the resistance of the coils. In a properly designed transformer, this loss of pressure should not exceed about 1 per cent. of the terminal pressure, so that the maximum difference in the pressure maintained in the secondary circuit between the lamps, when the apparatus is used for parallel distribution, does not exceed 2 per cent. between full load and no load, provided the pressure in the primary circuit is kept constant. Such a small variation in the pressure is well within the limits attainable in direct lighting, when a compound-wound dynamo is driven from an engine with a fairly good governor, and the corresponding variation in the illuminating power of the lamps is sufficiently small to be tolerated. A greater variation, especially if it occur suddenly, would, however, not be permissible; and the question, therefore, arises as to what rules should be followed in designing transformers so that they may as nearly as possible be self-regulating. The obvious answer is: make the resistance of the circuits as small as possible, and this may be accomplished in three ways: 1. By using wire of large area. This expedient naturally increases the weight, bulk, and cost of the transformer, and must, on account of commercial reasons, not be pushed too far. 2. By employing a core approximating to a circular or quadratic section, so as to decrease the perimeter of the coils. 3. By working with a high induction or with a high frequency.

The mean electromotive force in volts, induced in the coil of a transformer, can be expressed by the following formula—

$$e = (q)\, 2\, \pi\, n\, z\, \tau\, 10^{-8} \quad \ldots \ldots \quad (11),$$

in which (q) is a coefficient denoting the ratio between the maximum and the mean electromotive force (when the current-wave may be taken to follow a sine-function, this coefficient is $\frac{1}{\sqrt{2}}$), n is the frequency, z the total induction in absolute units, and τ the number of turns contained in the coil. The larger z, and the smaller n, may be the length of wire, and therefore its resistance. On purely theoretical grounds it would thus seem that a small core strongly magnetized, and as high a frequency as the alternator can be constructed to give, would produce the best results. In practice, however, this conclusion is wrong, and the heating of the iron core in consequence of hysteresis must be taken into consideration. Since a transformer is an apparatus at rest, and is frequently placed in a position where the cooling effect of the air is small, the ratio between heat generated and available cooling surface must be very different from that permissible in a dynamo, and it is this ratio which is the true limit to the output of a transformer. The armature of a dynamo is not subjected to excessive heating if a cooling surface of about 1 square inch per watt be provided. In a transformer, the amount of cooling surface should be at least three times, and if possible, four or five times as great, especially as it is somewhat difficult to determine, in a body of such complicated shape, what is and what is not effective cooling surface. From the table, giving the loss by hysteresis, it is easy to calculate for any induction, frequency, and weight of the active iron in a transformer, what the iron heat will be; and, adding to this the copper heat, the total cooling surface required to keep the apparatus at a safe temperature can at least approximately be determined, or the limit of output for any given transformer can be approximately found. In this connection, it is instructive to compare two similar transformers, one twice the linear dimensions of the other. For the purpose of the comparison, it may be assumed that in the small transformer the iron heat is the same as the copper heat, though it is generally greater. If the large transformer be worked at the same induction per unit area of core, its iron heat will be eight times that of the small transformer, or four times its total heat; but as the cooling surface is only four times as great, the anomalous result follows, that the output of the large transformer is zero, although its electromotive force would be four times that of the other. The comparison would have been still more unfavorable to the large transformer, had it been assumed that the iron heat in the small transformer was greater than its copper heat. Therefore in order to be able to work the large transformer at all, the induction must be reduced; that is to say, the transformer used on a circuit, the electromotive force of which is less than four times that of the small transformer. The resistance of the large coil is half that of the small one, and if the large transformer is worked so as to have equal iron and copper heat, the latter may obviously be four times that of the small transformer; this gives a current 2.82 times the value of that in the small transformer. The weight of iron in the large transformer being eight times that of the small one, the loss by hysteresis per unit of weight must be half. The corresponding induction can be ascertained from the table. If, for instance, the induction in the small transformer is 8,000, that in the large one may be taken as 5,000, and the electromotive force will be $\frac{5,000}{8,000} \times 4 =$ 2.5 times that of the small transformer, the output being increased in the ratio of 1 to $2.82 \times 2.5 = 7$ about. It should also be observed that the larger apparatus will, in so far, be more advantageous, as a smaller proportion of space will be occupied by insulation, especially if both are required for the same circuit. The ratio of output may thus be increased from seven to eight, or ten, or even more, according to the size of transformer taken as the basis of the comparison. For larger sizes, the advantage resulting from an increase of linear dimensions will be less apparent, and the output may roughly be taken as proportional to the weight.

The question of frequency, alluded to in connection with alternators, is also of considerable practical importance with regard to the transformers, and it is much to be desired that electrical engineers should adopt a common standard. Comparing two equal transformers, one working with a frequency of, say, 65, and the other with one of, say, 130, it is evident that with the same pressure the induction in the latter need only be half that in the former; and if the effect of static hysteresis only had to be considered, the heat generated per cycle would be something less than half; and the total heat generated in a given time would, with the high frequency, be somewhat less than with the low frequency. But as there is every reason to believe in the existence of viscous hysteresis, by virtue of which the heat generated per cycle increases with the speed at which the cycle is performed, there must be a limit to the frequency beyond which a further increase becomes disadvantageous. With the present imperfect knowledge of this subject, it is not possible to determine this limit on theoretical grounds; but it is justifiable to look to common practice for an indication of what the limit probably is. The present types of alternators and transformers, it must be remembered, are not mere experimental machines, but the survival of the fittest, and it would be futile to ignore the lessons taught by practical experience extending over several years. The following is a list giving the average frequencies adopted by several designers: Ferranti, 67; Lowrie-Parker, 88; Mordey, 100; Zipernowsky, 42; Kennedy, 60; Kapp, 80; Westinghouse, 133.

Taking the average of European practice, the frequency is 73, or 8,750 reversals per minute; whilst the American practice is 16,000 reversals per minute. This large difference may, to a certain extent, be accounted for by the general preference of European engineers for machines running at moderate speeds: but even when making due allowance for this circumstance, it can hardly be assumed that, had actual practice shown the value of a very high frequency, electrical engineers would not have found means to obtain it. As a matter of fact, most makers of alternators have started with a very much higher frequency than they have finally adopted. It should also be remembered that the higher the frequency, the greater is the loss of conductivity in the mains, owing to the unequal distribution of current throughout their cross-section.

Motors.

Machines, for the conversion of the electrical energy of alternating currents into mechanical energy, have not yet been brought to such a state of perfection that the problem may be considered as solved—and, in fact, very little is at present known as to the details of such machines and the results achieved. Motive power may be obtained from an alternating current in either of three distinct methods:

1. By the employment of an alternator with separately excited field, and some means to first run it up to speed before switching the current on. Various de-

vices have been patented for this method of working; but the author is not aware that any one of them has proved successful.

2. By the employment of a direct-current motor with laminated field-magnets. Such motors have been made and tested by various engineers; and about two years ago the author also experimented with such a motor, but the results were discouraging. The work obtainable from this motor, with a given current and electromotive force, was only about one-fifth part that which might be obtained from the same motor if driven by a direct-current of the same measured strength, and flowing under the same pressure. This particular type of motor was, therefore, commercially, quite unfit for use with an alternating-current supply. Afterwards, when investigating the reason of this failure, the author found that the electromotive force of self-induction was far too great, in comparison to the counter electromotive force of the armature, producing thus a very large lag, and conse-

quently reducing the plant-efficiency too much. The investigation also showed that, under the most favorable conditions, the plant-efficiency of such a motor could not be more than 70 per cent. This figure would be obtained if the electromotive force of self-induction of the whole machine were by some means reduced, so much as to be only equal to the counter electromotive force developed in the armature, a condition which is extremely difficult, if not impossible, to fulfill in practice. For this reason, it would seem that a motor of this type must always be very much larger, more costly, and heavier than a direct-current motor of the same power. It has the advantage of being self-starting, and not requiring any accessory apparatus for its excitation or regulation. On the other hand, there is the liability of the armature burning up, if the motor should fail to overcome its load almost at once. When the armature is at rest, and an alternating current passes through the machine, those armature coils which at the time happen to be short-circuited by the brushes, are in the same condition as the secondary of a transformer short-circuited upon itself. They will therefore be liable to burn up if the armature cannot start at once.

3. The third method of producing motive power from alternating currents has hitherto received most attention, as being the most promising. It is due to a discovery made about a year and a half ago by Professor Galileo Ferraris, of Turin. This scientist found that a copper cylinder, suspended between two coils, was set into rotation if alternating currents of the same period, but of different phase, were sent through these coils which were placed at right angles to each other. The explanation, given by Professor Ferraris, was that the resultant field of the two currents revolved round the common center-line of the coils, and, by means of eddy currents created in the copper cylinder, dragged the latter after it. This principle has been practically developed by Mr. Nicola Tesla and others, and motors have been actually built in

which a revolving field causes, by a kind of electro-magnetic drag, an armature to revolve and give off mechanical work. This action may be represented by the diagram Fig. 12, where A A and B B are two coils placed at right angles, and in

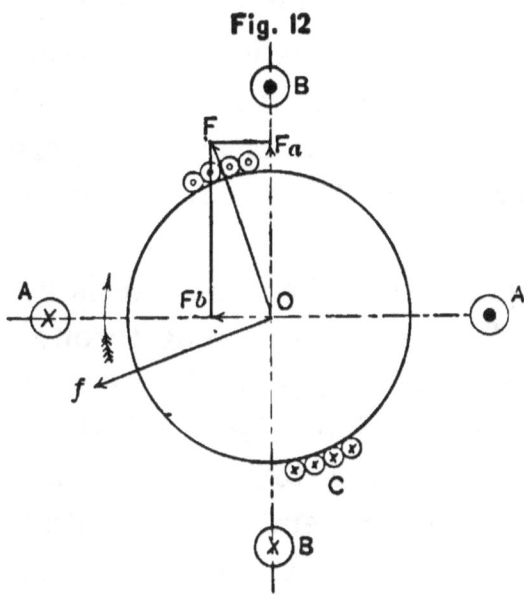

Fig. 12

the common center, O, of which is placed an armature consisting of a series of coils C, each forming a loop closed on itself. Let there be a difference of phase of 90° between the currents in the two field-coils, and let OF_a and OF_b represent the fields produced at a given moment. The resultant field is O F, and this revolves round O with a speed corresponding to the frequency. In the diagram a cross placed in the circle, representing a wire or coil, signifies an ascending, and a dot a descending current. When the rotation of the resultant field takes place in the direction of the arrow, the current in B must be approaching zero, and that in A its maximum value.

Suppose, now, the armature is held at rest. The field in sweeping through the conductors C induces in them currents in the direction indicated, and the latter therefore exert a torque. Imagine the field stationary, and the armature revolved by a belt. In this case, the current created in the short-circuited coils will produce a torque tending to resist rotation. The lower the resistance of these coils and the greater the speed of rotation, the more power will be required to rotate the armature. It is thus evident that with a revolving field the greatest force will be exerted upon the armature whilst this is at rest, and that the power will decrease as the armature begins to revolve and gather speed, just as in an ordinary direct-current motor. When the speed of the armature coincides with that of the resultant field, no torque is produced, whilst at starting, the torque is a maximum. According to the load put upon the motor, the armature will therefore automatically assume that speed which corresponds to the load, and in this respect the condition of working is comparable to that of a continuous-current motor. The current through the armature coils C produces a field, Of, which is also revolving, and in sweeping past the wires in the fixed coils A A, B B, creates therein an electromotive force which is always opposing the flow of current, as will easily be seen from the diagram. It will also be observed that the wave of counter electromotive force in the fixed coils, thus produced, coincides with the current wave therein, and the current must therefore do work in flowing against this counter electromotive force. It is this work thus absorbed by the fixed coils which partly reappears as mechanical work of rotation given out by the armature, the rest being dissipated in heat as in the case of an ordinary motor. The action here explained is in reality not quite so simple, because self-induction and the reaction of the armature field have to be taken into account; but, in the absence of all experimental data, it is not worth while to attempt a closer investigation. The author has thought it expedient to explain the fundamental principle of this type of motor, as the investigation might prove useful to those who are about to experiment with similar machines. He regrets that he is unable to place before the institution details of the Tesla motor, as constructed by the Westinghouse company; but as these details are only partially worked out, and secured by patents, this company naturally prefer to withhold them for the present.

Description of Modern Alternators.

By the courtesy of the designers and makers of alternators, the author is able to submit particulars of the various machines most in use, and he takes this opportunity of tendering his thanks to those who have supplied him with information. Modern alternators may be divided into two classes, disk machines and drum machines. In the former type the armature-coils are arranged in disk form, and the magnet-poles are presented to them from two sides, whilst in the latter type the coils are arranged to form a cylindrical surface, and the magnet-poles are presented from one side only.

DISK MACHINES.

The Ferranti Alternator.

The magnet-cores are of trapezoidal section (Figs 13 and 14), and supported in cast-iron yoke-rings. The armature-conductor is a thin corrugated copper strip, wound with a strip of vulcanized fiber of equal width upon a brass core.

To avoid eddy currents in the latter, it is subdivided into corrugated narrow strips, separated from each other by asbestos.

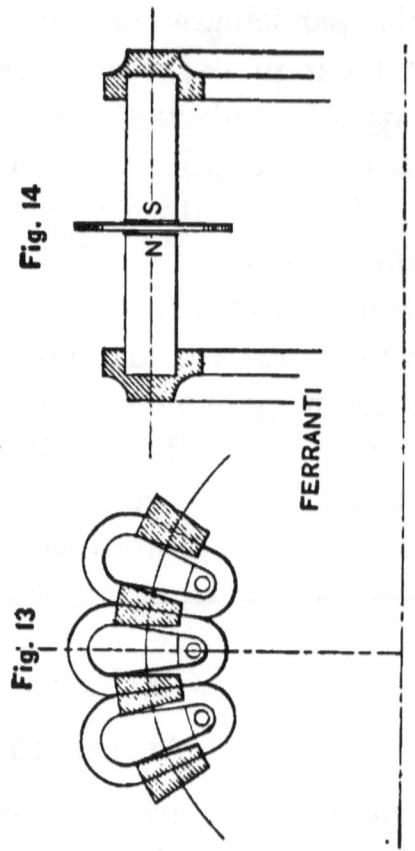

The shape of the coils is shown in Fig. 13. At the inner and narrower end, the core is provided with a projection by which it is secured to a bobbin-carrier.

This in turn is supported by, but insulated from, the central armature-ring. There are two bobbins to each carrier, which forms the metallic connection between their inner ends. There is thus a double insulation for the coil, first that of the conductor against the core and bobbin-carrier; and secondly, that of the latter against the body of the machine. The winding of the coils is such, that the current passes through the armature in two parallel circuits, by which arrangement the terminals of the armature are brought to opposite points of a diameter; and the connection between adjacent coils is made automatically, by the act of inserting and bolting up the coil so that it is impossible for a workman, when replacing a coil, to make a wrong connection. The mean width of poles is approximately equal to half the pitch, and the width of coils is also equal to half the pitch, so that, according to the table on p. 22, the coefficient is 2.300. The following are the principal data for a 150-HP. machine: 500 revolutions, 46.5 amperes,

2,400 volts, 20 poles. Armature-conductor 490 by 11.8 mils. Current density 4,000 amperes per square inch. Resistance of armature 1.2 ohm. Total number of conductors on armature 3,440, divided into two parallels of 1,720 each. Field-coils of 320 turns, exciting current 13.5 amperes. Exciting power 17,200 ampere turns. The interpolar space is 0.75 inch, and the area of pole-pieces 14.5 square inches. From these data it is possible to calculate the total flow of the lines. It is in English measure $z = 130$, and the electromotive force of an equivalent continuous current machine would be 1,150 volts. The ratio between this figure and the electromotive force of the alternator is 2.15 which agrees fairly well with the theoretical value of 2.30 mentioned above. The resistance of the field is 12.8 ohms, and the exciting energy 2,340 watts, or only 2.3 per cent. of the total output.

The Mordey Alternator.

This is also a disk machine; but with revolving magnets and fixed armature

(Figs. 15 and 16). The cores of the armature-coils are of porcelain, and the conductor consists, as in the Ferranti machine, of a thin copper strip insulated in a similar manner. The distinguishing feature of this machine is the field. It is of the iron-clad description, and the poles on one side are all of the same sign, whilst those on the other side of the

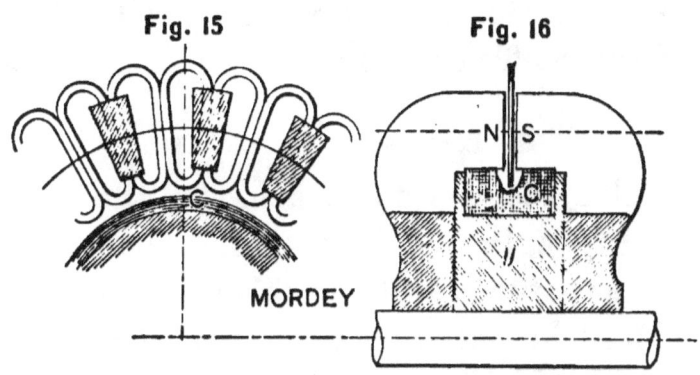

armature are of the opposite sign. There is only one coil of exciting wire, C, placed as shown in the diagram. The field may be diagrammatically represented by Fig. 4. Each group of conductors is fairly concentrated, and the coefficient k for this machine should, therefore, lie between the values given for cases 1 and 3 on p. 22. If the conductors were con centrated into lines, k would be equal to 2 000; if they occupied half the available space, it would be 1.635. As will be seen from the diagram, they occupy rather less than half the available space, and k should, therefore, be greater than 1.635, though smaller than 2. The author has not been able to obtain the constructive data of this machine, and can, therefore, not give the exact value of the coefficient k for it.

THE KENNEDY ALTERNATOR.

This, too, is a disk machine of the ironclad type, the field being stationary and the armature revolving (Figs. 17, 18, and 19). The latter has an iron core, A, round which the armature-conductor is wound, forming radial coils with parallel sides. The field-magnet consists of an external yoke-ring y bounded by flat annular disks, which on their inner faces are provided with polar projections N S, occupying relatively to each other intermediate angular positions. There is only one coil of exciting wire, which is, however, for greater convenience of manufacture, arranged in two halves, C C. Fig. 19 shows diagrammatically the relative position of armature-coils and field-poles straightened out. It will be seen that the lines of force, which are shown dotted, have an oblique course, and that only one-half of the wire in the armature-coil is at any time in the position of greatest activity. The width of the field, when making due allowance for the fringe of lines which surrounds the actual contour of the pole-piece, is somewhat greater than the pitch; whilst the proportion of space on the armature, occupied by the winding, varies with the radial distance at which it is measured. On the inside of the armature nearly the whole space is so filled, whilst on the outside a little less than half the space is occupied by the coils. For the outside of the armature the coefficient k will, therefore, be nearly that corresponding to case 3 on p. 22, and for the inside it will be rather less than that of case 2. The actual mean value of the coefficient should therefore be a little greater than 1.000; and this is, indeed, the case. According to a calculation made by Mr. Kennedy, the number of English lines emanating from one pole is, for the machine illustrated in the diagram, 148; the armature contains twelve coils of 18 turns each, or 216 turns in all; the exciting-coils con-

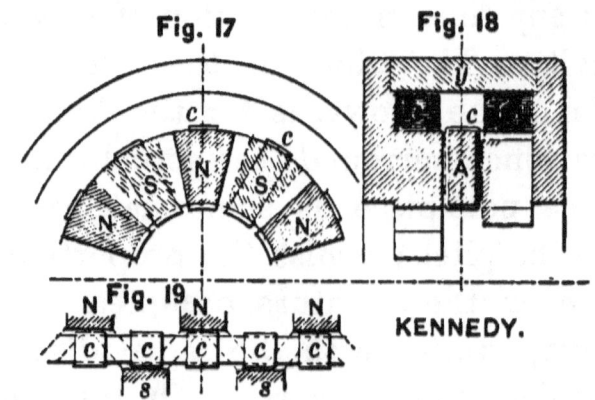

tain 570 turns of wire, the exciting-current is 25 amperes, and the exciting-power is, therefore, 14,250 ampere turns. At a speed of 800 revolutions per minute, the terminal pressure is 150 volts, and the current 70 amperes, the output being thus 10.5 kilowatts. No experiments have been made to determine the induced electromotive power, which naturally is effected by the self-induction of the armature; but Mr. Kennedy states that the difference in exciting-power for full output and running idle is not great. From analogy, with his own machines, the author estimates that the induced electromotive force of this machine, when giving 150 volts at the terminals, would be about 165 volts. If the armature were coupled to give a continuous current, the electromotive force produced by a field of 148 lines would be 154, so that k in this machine must be about 1.07. This value, although rather smaller than theory would indicate, is yet within the limits given above. It may here be mentioned that Mr. Kennedy advocates the use of low-pressure alternators in connection with step-up transformers if a high-pressure current is required.

The Kapp Alternator.

The author's machine is illustrated in Figs. 20 and 21. The armature-core A consists of thin band-iron coiled upon a cast-iron supporting-ring, and surrounded by radial coils c. The yoke-rings y are of cast-iron and the magnet-cores of wrought or cast-iron with pole-shoes of rectangular shape. This arrangement has been adopted with a view to obtain, on the outside of the armature, as large a ratio as possible between the pitch and length of field. For the same purpose, and to reduce leakage, the pole-shoes are tapered on the inside. The importance of securely supporting the armature-wires is well known, and there is hardly any good continuous-current dynamo now made, in which this mechanical detail does not receive careful attention. In alternators, a strong mechanical fastening of the wires on the armature is, however, of even greater importance. The mechanical strain to which the wire is subjected depends, not on the mean current and electromotive force at which the machine is rated, but on their maximum values. Now the maximum value of the periodic current is about 40 per cent. greater than the mean value, and the same is the case with the electromotive force. It should also be observed that the strain depends upon the induced electromotive force, whereas the output is computed on the basis of the terminal electromotive force, which is somewhat smaller. As the crest of the current-wave occurs in ordinary work only a very short time after that of induced electromotive force, the mechanical strain to which each coil is subjected twice in each period is that due to the maxima, and not to the mean values of current and electromotive force, and is therefore, roughly speaking, at least twice as great as in a continuous-current machine of the same dimensions and output. To provide against these strains, the author inserts between the coils driving-horns of insulated metal or fiber, which pass through the core of the armature, and are secured in their position by radial bolts. The machine illustrated in Figs. 20 and 21 is one of five 120-kilowatt alternators now in course of construction for a central station; but as it has not yet been tested, it will be more satisfactory to give the constructive data and results of a smaller machine which has been at work. For this purpose may be selected the machine to which the characteristic curves, Fig. 11, refer, and which has been exhaustively tested by Mr. Brown, as already mentioned. The supporting-ring of this machine is 31 inches in diameter and 3 inches wide. The core is of equal width

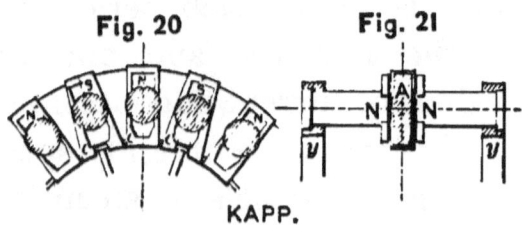

Fig. 20   Fig. 21

KAPP.

and $8\frac{1}{2}$ inches deep, and is wound with 1,120 turns of wire arranged in fourteen coils of 80 turns each. The resistance of the armature when warm is 1.74 ohm. The field-magnet cores are of cast-iron, 3 inches by 7 inches, with rounded corners, and there are no pole-shoes. Each magnet-coil contains 140 turns of wire, and the total resistance of the twenty-eight coils is 1.73 ohm. The calculated strength of field at 6,000 ampere turns is 142 English lines, and the observed terminal pressure when running idle is 1,560 volts. This gives a coefficient of $k = 2.370$. When working at the full output of 30 amperes and 2,000 volts, the mean electromotive force of self induction is 1,250 volts, and the induced electromotive force 2,400 volts. The loss of pressure through armature-resistance is in that case 52 volts, or about 2.6 per cent. of the terminal electromotive force. The energy lost in exciting the field, which requires 12,000 ampere turns, is 3.2 kilowatts, or 5.3 per cent. of the output. This somewhat large percentage is due to the employment of cast-iron for the magnet-cores, and will, in the machine illustrated, which has cylindrical magnets of wrought-iron, be reduced to about 3.5 per cent. The data, obtained by Mr. Brown in testing this machine, suffice for the determination of the coefficient of self-induction by means of a diagram similar to Fig. 7, which also gives the angle of lag. Knowing the latter, it is possible to find the counter-magnetomotive force due to the armature current at the moment when the armature-coils are in mid position between two poles. The maximum current is $30 \sqrt{2} = 42.4$ amperes, and the corresponding exciting power is 3,400 ampere turns. This, however, occurs at a time when the coils are well within the polar surfaces of the field-magnets and the poles produced in the armature are between neighboring field-poles. In this position the demagnetizing effect must be naturally small. It will be a maximum when the armature-poles are opposite the field-poles, and at that moment the exciting power of the armature-current is about 1,700 ampere turns. This value must, therefore, be deducted from the exciting power applied to the field, to obtain the true effective exciting power which is causing the flow of lines. Thus the value 10,300 is obtained, to which corresponds a flow of 215 lines through the armature, giving the induced electromotive force of 2,400 volts.

DRUM MACHINES.

The distinctive feature of the machines comprised under this title is that the armature-coils do not surround the core, but are arranged on one side of it. As examples, may be mentioned the Westinghouse, the Lowrie-Parker, and the Zipernowsky machines.

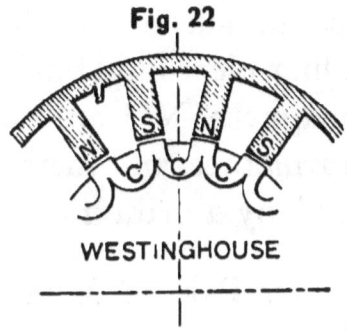

Fig. 22

WESTINGHOUSE

THE WESTINGHOUSE ALTERNATOR.

The armature-core of this machine is of cylindrical shape, and consists of thin iron plates, as in an ordinary drum machine for continuous currents. To provide internal cooling surfaces, the plates have large holes cut out of their surface, forming, when put together, tunnels, going from end to end, through which the air may circulate. The armature-coils are of flat-link shape with rounded ends, which are laid on to the surface of the core with the active wires parallel to the spindle. The ends of the coils are brought over the end plates of the core, and bent inwards, as shown at C C C in Fig. 22, and secured in that position. The straight portions of the coils are fastened to the surface of the core by insulated filling-pieces and binding-hoops. The field magnets N S are set radially outside the armature, and their outer ends are connected by a cylindrical yoke y.

The Lowrie-Parker Alternator.

This machine is the inverse of that just described. The armature is outside of the field, and is stationary, whilst the magnets revolve. The armature coils C C, Fig. 23, are also of flat-link shape, the conductor consisting of a cotton-covered strip, wound upon a core of wood with rounded ends; they are laid with their long side parallel to the spindle on to the inner surface of the armature-core

A, which is built up of segmental plates to form a cylinder. The following particulars refer to a 100-kilowatt machine, giving 50 amperes at 2,000 volts terminal pressure, the speed being 380 revolutions per minute. The internal diameter of the armature core is 5 feet 1 inch; the radial depth 5 inches, and the length 14

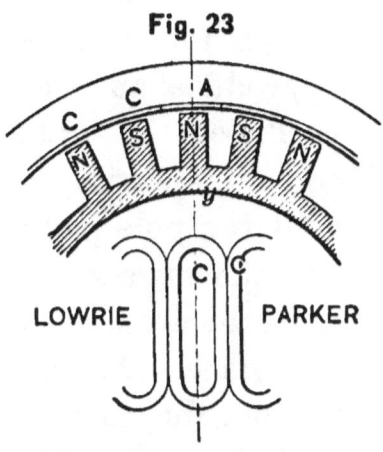

Fig. 23

inches. There are twenty-eight coils, each containing 21 turns, so that the total number of active wires, counted all round, is eleven hundred and seventy-six. The field-magnet cores are of wrought-iron, 3 inches by 13½ inches, and are bolted to a wrought-iron yoke-ring y. Each magnet-coil contains 130 turns of exciting wire, and the exciting current is 28 amperes. This gives an exciting power of 7,300 ampere turns, and the calculated strength of field is 144 English lines. Allowing for self-induction and resistance, the induced electromotive force may be taken as 2,050 volts. A continuous-current machine of the same dimensions and winding would give 900 volts; so that for this machine, as arranged to produce an alternating current, the coefficient is $k = 2.280$. The width of the field, allowing for the fringe of lines round the pole pieces, is 0.55 of the pitch, and the surface covered by the active wires in the armature-coils is 0.55 of the whole surface. Theory would therefore indicate that the coefficient must be very nearly that given for case 5, p. 22, namely, 2.300, which is indeed the case. In a 150-kilowatt machine of the same type, the output is 75 amperes at 2,000 volts pressure; the speed being 500 revolutions per minute, and the frequency 100. There are twenty-four poles, and the number of active wires is six hundred and twenty-four. Resistance of armature, 0.32 ohm, and that of the magnets 25 ohms; exciting current, 15 amperes.

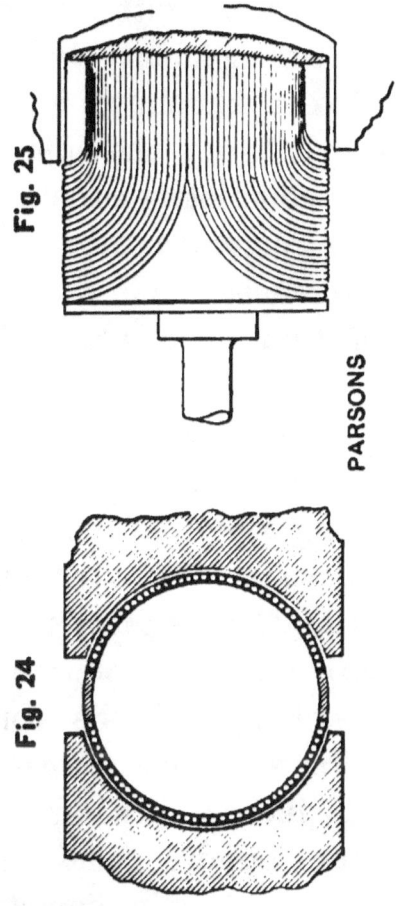

THE PARSONS ALTERNATOR.

This machine, of which Fig. 24 is a

section, and Fig. 25 an elevation, is a modification of the well-known turbo-electric generator of the Hon. Charles Parsons. It is the only example of a bi-polar alternator with drum armature, the arrangement being intermediate between that shown in Fig. 4 and Fig. 5, that is to say the polar surfaces, although not completely surrounding the armature, yet embrace more than half the circumference. The armature core consists of thin iron plates insulated with paper and mica, and threaded upon the spindle. The wire is laid on in a single-coil, the ends being brought to two collecting rings in the usual way. In a 75-kilowatt alternator, giving 75 amperes at 1,000 volts terminal pressure, the armature core is 7 inches in diameter by 30 inches long, and the core of the field-magnets, which are of cast-iron, is 5 inches by 28 inches. The armature coil contains one hundred and ten active wires, 127 millimeters in diameter, giving a current-density of 6,000 amperes per square inch. The armature resistance is

0.22 ohm, and the loss 1.65 per cent. of the output. The speed of this machine is 6,000 revolutions a minute, giving a frequency of $n = 100$. The armature weighs 360 lbs., and the weight of the complete machine, including the turbo-motor, is about 2 tons.

The Zipernowsky Alternator.

This machine is very similar to the one just described, the principal difference being that, instead of a smooth core, the armature is provided with a core having Pacinotti projections on the inside. The magnet cores N S, Fig. 26, are composed of U-shaped iron plates, to avoid the heating which would otherwise result from the employment of an armature with projections. The core of the armature is formed of thin sheet-iron segments A, each resembling a very short or shallow T, the armature-coil C being placed over the central stem of the T. Each armature-segment with its coil forms a separate part, which can be inserted or withdrawn without disturbing the rest of the machine. This is a great practical advantage, because facilitating repairs. There is a slight increase in magnetic resistance due to the want of continuity of armature-core, but, as the interpolar space can be made very small, the increase of magnetic resistance in the armature-core is permissible.

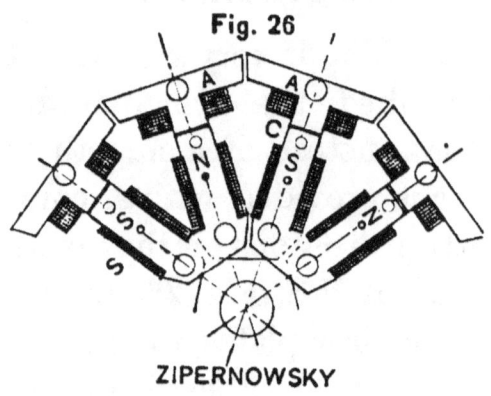

Fig. 26

ZIPERNOWSKY

The following particulars refer to a machine giving 40 amperes at 2,000 volts terminal pressure, or 80 kilowatts output. There are fourteen poles in the field, and fourteen coils in the armature. The speed is 360 revolutions per minute, and the frequency is $n = 42$. The weight of laminated iron in field and armature collectively is 1 ton 7 cwt., and that of copper in the armature and field-coils collectively is 930 lbs. The armature resistance is 1.038 ohm, and that of the field coils is 3.24 ohms; exciting current, 28.7 amperes. The loss of energy in the armature is thus 2.08 per cent., and that in the field 3.33 per cent. of the output. Some experiments were made with this machine to determine the loss due to windage, mechanical friction, and magnetic friction. For this purpose the machine was driven by a belt from an electromotor, the efficiency of which had previously been determined. For the first experiment, the field of the alternator was not excited, and the power required to drive it, at its normal speed of 360 revolutions per minute, was found to be 4.07 HP. In the next experiment the field was excited so as to produce a terminal pressure of 2,000 volts; but no current was allowed to flow through the armature, and the power required was found to be 9.81 HP. From these experiments it appears that the loss through magnetic friction was 5.74 HP. The total commercial efficiency at full load, when the alternator is driven by a belt, and inclusive of the power required for exciting, is given by the makers as 87 per cent. When driven direct by a high-speed engine, the efficiency is slightly greater, as there is no increase of journal friction by reason of the strain in the belt.

DISCUSSION.

Sir George B. Bruce, President, said the paper was evidently the work of a perfect master of the subject of which he treated, and he was sure that the members would award the author a vote of thanks for so valuable a contribution; nor did he think that he was sanguine in anticipating that it would give rise to a very important discussion.

Mr. Gisbert Kapp was afraid the members had found the paper somewhat dry; but he had asked several friends to enable him to render the proceedings a little less dry by lending him some apparatus. Most of the exhibits were new, and were then shown for the first time. Amongst the transformers was Mr. Mordey's, consisting of rectangular iron plates and bridge-pieces, the long rectangular coils being threaded through the cavities thus formed. The construction was very simple. The metal stamped out of the rectangle formed a bridge-piece, thus completing the magnetic circuit. The transformers were built up in such lengths as would give the required pressure, and the same plates were used for various sizes. If half the pressure was wanted, half the number of plates were used. There were other transformers exhibited, two by Messrs. Lowrie and Hall, which were used in Eastbourne, and were about to be used in London, and also two of the author's type. The core of the Lowrie-Hall transformer was composed of iron plates insulated with varnished gauze, and brought together over the ends, where they were protected by iron caps. The coils were wound on both limbs. There was a 20-HP. transformer of his own of last year's pattern, made by Messrs. Goolden and Co., and another, the latest pattern, made by Messrs. Johnson and Phillips. The difference between them was that one was adapted for indoor and the other for outdoor use. They could act under water, and had been so worked for trial, but of course no one would think of working them in that way permanently. In the new pattern transformer the winding was split up, so that the difference of potential of any particular coil would only be a portion of the total pressure. The arrangement had another advantage. If transformers were wanted for two different voltages, say 50 or 100, the same instrument could be used, as the two secondary coils could be coupled in parallel for the lower and in series for the higher pressure. Some iron tubes and steel boxes were shown, as used by Messrs. Lowrie and Hall for distribution by underground mains, drawn through these tubes. There were also samples of lead cables, the joints of which were made in an extremely ingenious way. The lead was first stripped off, and the insulation removed over a distance sufficient for making the joint; when that had been done the insulation was replaced, a plate of lead was put over it, and a tool like a pair of nippers took it round and squeezed it into the shape shown, making a very strong joint. The connections were made in the boxes, which were of three different types. Among the exhibits was one of Mr. Lowrie's meters, containing a secondary battery in series, with a set of copper depositing plates and the consumers' main leads. It did not measure currents, but conductivity-hours. If the lamps were switched on, but no alternate current was sent through, and the circuit was simply completed, the meter registered conductivity-hours; but as the pressure was always on the mains, that was equivalent to registering energy. It had been in use for two years with good results. There was also an ampere meter for alternate currents. The current was measured by the heating effect upon compound strips of metal. The heat caused the strips to deflect, and the motion was recorded by a pointer. There was also shown a safety fuse which would be melted, and thus interrupt the current when it exceeded a predetermined limit. With a pressure of 2,000 volts it was necessary to have a long fuse-wire, because the arc set up upon the melting of the fuse would remain, unless its length were considerable. As a further precaution to make the arc cease, the wire itself was within a slot in the slate base, and the arc was thus subjected to the chilling effect of the cold slate. He was able to exhibit a portion of a switch of Mr. Lowrie's installation in Kensington, by which the connection between the different dynamos and circuits could be effected. Also an extremely neat arrangement for showing the insulation of the cables. It was nothing more than a Geissler tube. One pole was put to earth, and the other was put to any one of the circuits which were to be tested. If the insulation was good on that particular circuit, the tube showed a bluish light, which was observed through the eye-hole. If it was faulty, this static effect did not take place; the tube remained dark, and the attendant knew that the particular main was faulty, and must be repaired. If complicated tests were used, requiring elaborate apparatus which could not be easily carried about, the men would be sure to neglect it. With a little instrument like that exhibited, two or three circuits could be tested in half a minute. The author also exhibited an armature-coil of one of Mr. Mordey's dynamos shown in Fig. 15; also various measuring instruments, a safety apparatus invented by Captain Cardew, and various photographs of the large dynamos made by Mr. Ferranti.

Mr. W. H. Preece said that it might not be out of place if he were to give a brief account of the steps, taken in recent years, to solve the very difficult question of the economical distribution of electricity over large areas. The paper was, as the author had himself said, dry; but it really recounted the gigantic advances that had been made in the means required for this economical distribution of electricity. It was only ten years since Edison solved the question of incandescent lamps, and only seven years ago it was found scarcely possible to distribute currents over areas so as to bring electric lighting to compare in any way with gas. At that time a distinguished French electrician, Mr. Gaulard, showed how, by using alternate currents of high electromotive force, to distribute electrical energy to a distance by stepping down, as it were, from high pressure to low pressure. The process was similar to that now carried out in London, to distribute power by means of water under very high pressure; also to that adopted by the Gas Light and Coke Company at Beckton, to send gas under high pressure to holders in different parts of London, and there distribute it under low pressure. Such progress had been made in America, on the Continent, and in England, that ere very long he believed there would be electric lighting all over the country. There was hardly a town of any size that was not considering the subject, and that, in twelve months' time, would not be on the road to having electric lighting distributed throughout large areas. At Deptford, there was the extensive enterprise of the London Electrical Supply Corporation, and the enormous machines of Ferranti were being constructed, intended to produce an electric pressure equal to 10,000 volts, representing an electromotive force something like that of a small lightning flash. It would be driven through a very small conductor up to the neighborhood of the Monument, and would there step down to a lower pressure, and so be distributed throughout the city of London over a very large area. One was the proper rate at which the alternating currents should be transmitted. The practice had hitherto been to commence at a very high rate. Mr. Ferranti began with about 150 currents (300 positive and negative) per second, but by practice he had come down to the figure given by the author, 78. He had omitted to mention that throughout Europe the alternate-current machines were being constructed principally by the firm of Ganz, of Budapest. The Zipernowsky system had been introduced, first in Italy, and there worked out thoroughly by a distinguished electrician, Professor Ferraris. In the Zipernowsky system the frequency had been brought down from 150 currents to 42. Mr. Parker had brought it down to 100. That was one of the most important questions to be decided, and it could only be done by actual practice. Sir William Thomson had shown, by mathematical reasoning, that if currents were sent with high frequency, they had not time to penetrate into the interior of a conductor. With a solid conductor, say between Deptford and London, one inch thick, at a high frequency, the current would not enter more than about 3 millimeters inside the surface; thus the only part of the copper conductor conveying electricity was a thin shell or tube on the outside. If that theory was true, the lower the frequency, the greater the efficiency of the conductor. Another reason for less frequency was that meters could work with greater accuracy.

Professor George Forbes congratulated the author on the many points of interest he had brought forward, and especially on the happy way in which he had taken in hand the theoretical investigation of alternate-current machines. He had been one of the first to discuss the theory of continuous-current machines in the proper fashion, and since he read his paper on the subject before the Institution two years ago, the progress had been marked, so that now the design of continuous-current machines was as straightforward work as any mechanical designs. He had taken up the subject of alternate-current machines in the same spirit. On the occasion when the author read a paper on continuous-current machines, Professor Forbes had occasion to complain of the combination of different kinds of units, C. G. S. and English. He was sorry to find the same practice in the present paper, and he wished to enter his protest against it. The author spoke of the mean value of the electromotive force of an alternating current as the electromotive force of a direct current, which will, in a given resistance free from self-induction (say, for instance, the wire of a Cardew voltmeter), produce the same amount of heating. That was not a definition of the mean value of the electromotive force of the alternating current. He would beg the author to withdraw the expression "mean value," and substitute "virtual or equivalent value." Until a few years ago, electricians thought that a continuous current was the proper one to use for distribution, and that the alternate current was objectionable for many reasons. But now they were quite agreed that the alternate current was in many of its adaptations far more beautiful and more readily adapted than the continuous current. It was due to Messrs. Gaulard and Gibbs that a system of transformation from high to low pressure was economical, and could be adopted in central-station distribution. If any one doubted the value of the work done by those gentlemen, he would ask what would be the present position of electric lighting if Messrs. Gaulard and Gibbs had not, during two or three years of persistent opposition on the part of electrical engineers, forced upon electricians the conclusion that the use of their secondary generators was an economical mode of distributing electricity? While great credit was due to the author for having introduced his theoretical views on the question of alternators, it was unfortunate that he had not given more numerical facts as to alternating dynamos, as to the efficiency of the transformers, and so on. Such data were very desirable. The author had raised the question whether dynamos ought to be worked in parallel or separately. On the ordinary continuous system, as first generally adopted by Edison and now universally employed, it was customary to have all the mains in the district connected together into a large network, with feeding cables going to different points. Those feeders, all issuing from the central station, were connected to one set of mains, and all the dynamos were connected in parallel to those mains, all working together. There was a difficulty in alternate currents working in parallel, and it was customary not to make a single network over the whole of a district, but to subdivide it into small districts, each with its separate feeder, and each feeder might be fed by a separate machine or a number of the feeders might be grouped on to one machine. At the beginning of the evening's work all the feeders would be upon one machine. As the consumption of electricity increased, some of the feeders would be passed on to another machine, and so on. The author considered it impossible to switch the feeders on to a new dynamo without making some flickering in the light. In the course of last year Professor Forbes had examined a very large number of central stations, and the alleged difficulty did not exist. The facility of switching in dynamos without producing a flicker depended simply on the kind of switch employed. With a rapid switch there was no difficulty. He had lately examined the central station at Rome, set up by Messrs. Ganz and Company, and there was there a switch-board of very great ingenuity, designed by Mr. Blathy. As to the possibility of working conveniently in parallel he might say that experience in America had been completely against it. It was there found that it was possible to work in parallel, but that it enormously increased the amount of skilled attention required in a central station; and for that and other reasons it was far better to divide the district into a number of sub-districts, feeding by separate feeders from the station. That method had also another advantage, especially in a country like America, where overhead wires were chiefly employed. There was another reason why in America it had been preferred not to work in parallel, and that was that the alternations of current in the machines were more frequent, and such machines could not be worked in parallel so easily as those which had a small number of alternations. Dynamo machines for alternate currents might be divided into two classes, those which had little or no self-induction, and those which had large self-induction. The Siemens machine was one of those with little or no self-induction, and the Mordey machine was another of the same type. The armature of the Mordey machine looked very like that of the Siemens machine; but it was a fixture, and the Mordey machine had a peculiarity, in common with the Kennedy machine, which made it extremely original. In machines of the Siemens and Mordey types, where there was no self-induction, if it was desired to introduce it, all that had to be done was to put large self-induction into the circuit. This was as convenient as to put the self-induction into the machine itself. Still, there was a good deal to be said in favor of machines of iron. One advantage was that the clearance between the poles and the armature was smaller, and there-

Figs. 27

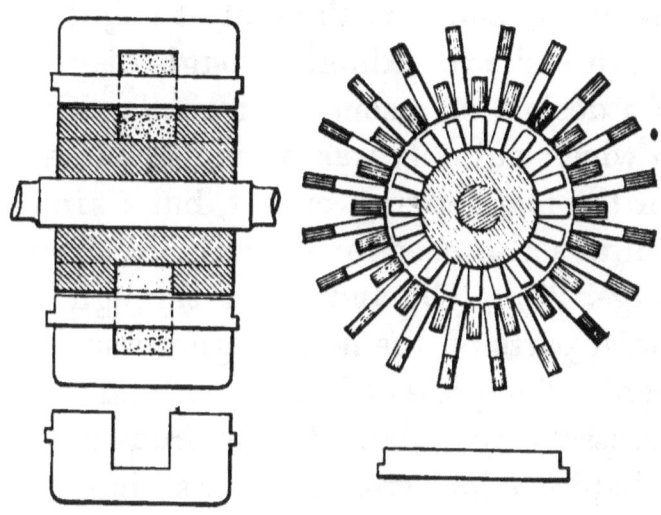

SINGLE-COIL ALTERNATOR.
Scale $\frac{1}{12}$.

fore the magnetization could be got at a cheaper rate. One of the great advances in alternators of late years was the invention of the Mordey machine. Up to that time alternate-current machines were made with a large number of pole-pieces. The Mordey machine had two; in the old type each pole-piece had its magnetizing coil, and there was great waste of energy. Mr. Mordey had introduced the idea of using a single coil to magnetize the whole machine. Professor Forbes had been experimenting a little in that direction, and had thought it might be worth while to exhibit a rough diagram of a design, of a somewhat similar kind, in which not only was a single coil used for the magnetizing circuit, but a single coil was also used for the induced circuit (Figs. 27). The induction through the radial parts of the field magnet was constant. To prevent heating, it might be necessary to subdivide the iron at the pole-pieces into circular sheets; but that would be seen when the first machine was made. That type of machine was certainly of interest, and might prove of value. The Parsons machine seemed to him to fulfill the requirements of an alternator better almost than anything else which had been produced. It also had the advantage of having a single current for the field magnet, and a single current for the induced armature. He would only say a word on the subject of the speed of alternations. Why was it that in America 16,000 alternations per minute were used, and in Europe at the greatest number of stations 5,000? One reason was, that in America high-speed machines were in favor, and in Europe low speed. Another reason was that, in America, the object aimed at was to get the greatest output from the plant. Those reasons were sufficient for having high-speed alternations. Unless it was wished to work in parallel, it could not be doubted that working with the highest-speed alternations was best; but for working in parallel there could be no doubt that the lowest speed was the most easy mode. That was one reason why the Ganz machines at Rome were so suitable. In this country there was a general belief that converters had an efficiency of 95 or 96 per cent. on full load; but the greater number of converters used in England had nothing like that efficiency. The smaller converters used by Messrs. Ganz and Co. certainly had nothing like it; they guaranteed an efficiency of 88 per cent. The machines exhibited, the Lowrie converter, the

Fig. 28

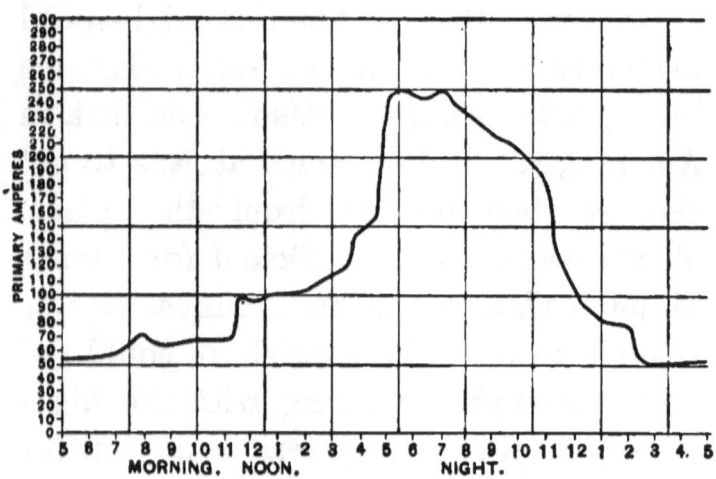

CURRENT OUTPUT. GROSVENOR GALLERY.

Kapp converter, and still more the Ferranti, had all a high magnetic resistance, which he was inclined to think would be unfavorable. He knew that tests had been made with a Kapp converter by Professor Ayrton, but in totally different conditions from the machines shown, and he should be surprised if they had the same efficiency as that shown by Professor Ayrton. He might take one example, which was very prominent, namely, the converters used by the Grosvenor Gallery. The diagram (Fig. 28) showed the output of amperes in a day, and it would be seen that between four and seven hours in the morning, when no lamps were on circuit, there was an indication of four thousand lamps being on circuit according to the amperemeters of the central station; that meant a waste of nearly four thousand lamps always going on, and with higher loads the loss was greater. The loss as shown in the diagram was 20 per cent. of the maximum with no load; consequently when there was any load on, the efficiency was certainly not 80 per cent.

Mr. Llewelyn Atkinson said that the constants which the author had developed, and which were very useful in comparing the electromotive force of alternating machines with machines of a continuous type, seemed to him might be made more useful if they were more separated. The constant K appeared to contain three things. First, it contained a multiplier of 2, introduced because alternate-current machines generally had only a single circuit, not two circuits, but this was not always the case, and the multiplier 2 should be put separately. Secondly, the constant included a quantity depending on the shape of the field and the arrangement of the armature; and thirdly, the quantity depending on the ratio of the square root of the mean square to the mean. All those three quantities appeared to be mixed up in one number, which might be useful in the particular cases given, but he thought that if they were dealt with as he suggested they would be more useful to subsequent designers. Nearly all cases which the author had treated so lucidly as to connecting machines in parallel, had been by the graphic method, which depended entirely upon the assumption that the electromotive force followed a law of simple harmonic motion. He was aware that there were great difficulties in treating it in any other way. First of all the law itself had to be found; and secondly, he did not know of any graphic method of solving the problems when the law had been found. The analytical method was probably even more difficult. It was generally complicated enough even with a simple harmonic law, but with irregular laws he did not know that it could be touched at all. He had therefore brought a diagram showing what were likely to be the errors even in the machines the author had treated of, taking the harmonic law to be true. In Figs. 29, the curves 1, 2, and 3 showed the primary electromotive force induced in a coil all wound in a single line on the armature, by moving it through three different forms of magnetic field, a harmonic field, a field such as occurred in the Mordey machines, and a field such as was given by the author's machine with opposite poles alternately on each side of the ring, with a space equal to the width of pole. 1. Gave a curve of primary current in a circuit with self-induction which was also harmonic, and a curve of secondary electromotive force also harmonic. 2. Gave a curve of primary current similar in form to the

Figs. 29

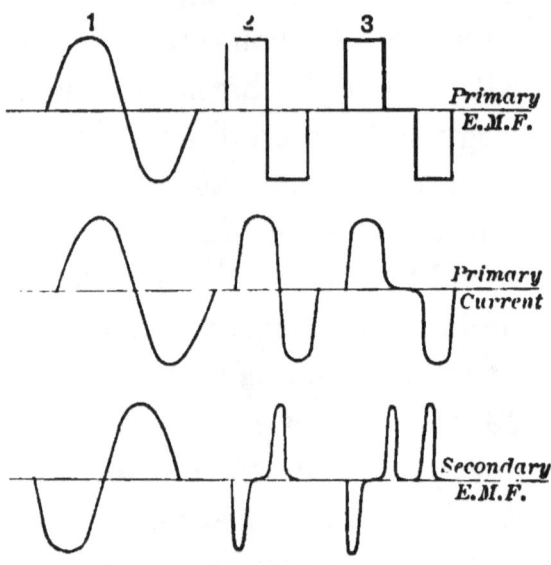

curve of electromotive force, but with all the corners rounded and the rises and falls less abrupt; but the curve of secondary electromotive force was different, and showed rapid rises and falls and intervals of no electromotive force.

3. Gave the same class of results as regarded primary electromotive force and current; but here also the curve of secondary electromotive force was very different. Owing to the fall in positive value of electromotive force being discontinuous with the rise in negative electromotive force, there were two impulses of secondary electromotive force, corresponding to each impulse of primary electromotive force. The effect of this was to double the loss from magnetic friction, or hysteresis, in any magnetic apparatus in the secondary circuit. That at once showed the very important bearing of the question of the law of electromotive force, and the errors that might be introduced by assuming a simple sine-function. He had alluded to the difficulty of treating the matter analytically and graphically; but there was one mode of treating it which he had not seen mentioned, the idea of which occurred to him two years ago, and he had designed some machines for carrying it out, namely, treating the subject by means

of the integrator (Figs. 30). To take the simplest case of a circuit, of which the induction and resistance were known, and

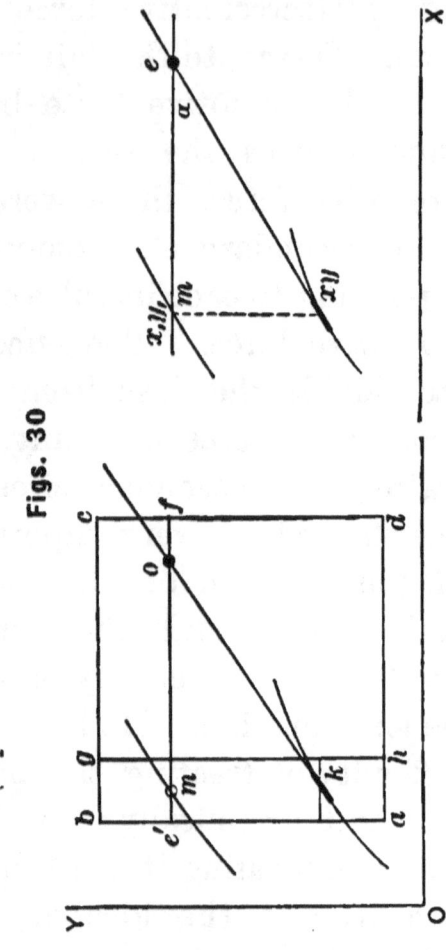

Figs. 30

the law of electromotive force was given by means of a curve. Let *a b c d* be an integraph in skeleton consisting of a rectangular frame. In this frame slid a cross-bar $e'\ f$, and within an inner slide $a\ b\ g\ h$, a carrier through which was pivoted, on a vertical axis, an integrating roller. The angular position of this roller was fixed by a rod $k\ e$, sliding at e through a pivot e, free to move round in a horizontal plane. The whole frame could only move in a direction parallel to the axis of x, being constrained by guides or rollers. Through the bar $e'\ f$ at m passed a pointer. Let $x,\ y,$ be co-ordinates of a point on the curve of primary electromotive force, let $x\ y$ be co-ordinates of the curve traced by the non-slipping integrator roller. Then by the construction of the apparatus the following relations held good. The roller was always tangential to the curve $x\ y$, and

$$\frac{d\ y}{d\ x} = \tan a = \frac{y_1 - y}{m\ e}.$$

Let the scale of y be such that $y = \dfrac{E}{R}$, and let $m\ e = \dfrac{L}{R}$, where L and R were coefficients of self-induction and resistance.

Then $\dfrac{dy}{dx} = \dfrac{\dfrac{E}{R} - y}{\dfrac{L}{R}}$, and if the co-ordinates $x\, y$ represented current and time respectively, $\dfrac{dc}{dt} = \dfrac{\dfrac{E}{R} - C}{\dfrac{L}{R}}$, or $L\dfrac{dc}{dt} + CR = E$, the well-known equation of a single electric circuit. Thus, by means of the integrator, any curve of electromotive force being given, it was possible to draw the curve of the current. He had further designed machines which it would take too long to describe, in which with two circuits, given the coefficients of self-induction, the coefficient of mutual induction, and the resistances, with a given curve representing the primary electromotive force, the curve of primary current, the curve of secondary electromotive force, and the curve of secondary current could be deduced; and all the curves, their shapes, their retardation, and everything about them.

Mr. James Swinburne considered that the author's papers gave rise to very good discussions, partly because they were of great value, partly because he always selected a subject which was fashionable at the moment, and partly because he always gave a good deal to disagree with. With regard to the author's assumption of the curve of sines, he could not see how that afforded any help. In practical work, the only use he had ever found in the curve of sines was in seeing whether an instrument, which had really a coefficient of self-induction, would give a reading within a certain percentage error. But in dealing with machines the assumption seemed to him to be an absurdity. It was only taken because it admitted of a great many mathematical calculations; but if the mathematical calculations were based upon empirical data he thought it was a pity to make them. Of course it was continually said that by Fourier's theorem any curve could be made up of curves of sines. In practical work electricians never dealt with anything that had constant coefficients of self-induction or mutual induction, except measuring instruments; and in dealing with transformers, incandescent lamps, and so on, nothing of the sort came in. He thought that all the comparisons in the paper between direct and alternate-current machines were based on an assumption which made them misleading. The assumption was that the direct machines had the same number of armature turns as the corresponding alternate machines. In practice he did not think that anybody ever made the air-gap otherwise than equal to the breadth of the pole. The output varied as the product of the number of useful wires in a coil, and the breadth of the pole-piece. The product was greatest when they were equal, and, consequently, designers always made them equal. The fringe of useful field at the edge of the pole-pieces must be allowed for; and in some cases the question was complicated by the sides of the coils being parallel, and the sides of the pole-pieces radial, as in the author's machine. The author assumed a constant coefficient of self-induction in the armature. In the machine which had originated the unwarrantable assumption of the curve of sines the self-induction was nearly constant. It was a Siemens machine used by Mr. Joubert. There was very little iron near the armature, and the coefficient of self-induction was sensibly constant. In the modern machine there was not a constant coefficient of self-induction in the armature. This would, he was afraid, detract from the usefulness of Mr. Atkinson's ingenious application of the integraph. The simplest way of calculating these things was not to take the coefficients of self-induction, but to find out how many lines of force were cutting the circuit at any time, by plotting the field-magnets out, and working through a graphic method. It was a long and tedious process, but it could be done, and done with the same accuracy as, for example, Professor Forbes' dynamo leakage calculations. In the direct-current dynamo the term self-induction had been a sort of scapegoat for all sorts of vague notions. Even now, in the case of direct-current machines, people talked about the self-induction of the armature. Of course there was no such thing in direct-current machines, and the idea was an absurdity. In alternating currents, the same accusation could not be brought, as the whole armature current varied. With regard to motors, he thought the author's ideas were a little elementary for the present day. With a big motor, running at different loads, it was found that the efficiency was extremely small if dependence was placed simply on self-induction to keep the motor and dynamo in step. He did not think that motors which depended on self-induction for keeping in step would answer. Many seemed to think the chief difficulty was to start the motor. This was really a small matter. All that had to be done was to make a self-exciting motor. Take the commutator machine made by the Thomson-

Houston Company, and put it on an alternate-current circuit, it would start, sparking badly at first, and eventually synchronize. There was no difficulty in starting the motor, but it would not work efficiently. The author had asked how it was that the waste power was greater in the armature conductors of a machine without iron in the armature. It was because the change of field was more abrupt in that case. A great deal had been said at different times about the advantage or disadvantage of iron in the armature, and it had been discussed both from a mechanic's and from an electrician's point of view; but when it was remembered that the air-space (in spite of what Professor Forbes had stated) was really the same whether there was iron in the armature or not, it would be seen that the whole question was one of mechanical construction against electrical efficiency. If there was iron, power would be lost in the reversal of magnetization. If there was none, there must be, to a certain extent, flimsiness.

There might be a more efficient machine, but there was mechanical flimsiness. A great deal had been said about hysteresis. He considered the author's inductions from Professor Ewing's paper were faulty. Plotting out the curve from that paper, or from Dr. Hopkinson's paper, it would be found that up to an induction of about 16,000, the power spent with a given alternation frequency varied sensibly as the induction. The author's argument, that if there was half the induction there was half the heat per cubic centimeter, was all very well, but then twice the iron must be used to get the same output. And it was found that in transformers made according to present practice with low inductions, in which most of the heat was wasted in the iron, as it was in transformers of any reasonable size, the speed of alternation made very little difference theoretically in the output. He said theoretically, because not enough was known about the time lag, and if that was great, larger outputs could be obtained with few alternations.

He was aware that he was disagreeing with Professor Forbes on that point. He hoped Mr. Atkinson would be able to supply some particulars as to time lag, that he would run the machine at different speeds and give the results. The kind of time lag mentioned by Professor Ewing was well known to dynamo makers, and the phenomenon was familiarly known as "creeping up." A new machine sometimes took twenty minutes to reach its full excitation. Messrs. Crompton & Company were making experiments on the important question of loss of power in iron in alternate-current work, which he hoped would be published. The question of hysteresis concerned transformers as well as dynamos. He thought that most people used bad iron. In testing iron for direct-current machines they only thought whether, with a given excitation, they could get iron highly magnetized. That was not the important question in alternating currents. The question of hysteresis must be considered. There was a great difference in samples in that respect. He had made transformers which would work with very high induction without heating. One of the most important questions to settle was frequency of alternation. It had been stated that in America 16,000 alternations per minute had been taken, and at Rome 5,000, and what should be made the standard in England should be carefully considered. In dynamo construction he found that it was easier to make a machine giving a very few alternations. In that way he could construct a larger dynamo for the same amount of money, and make it work more efficiently. The first point to be settled was the number of vibrations per second which could be separated by the eye. Helmholtz, in his "Sensations of Tone as a basis for the Theory of Music," gave 24 per second. Messrs. Crompton were going to try some experiments to ascertain whether it was possible to detect flickering in 100-volt lamps with anything below 24 vibrations per second. It was a pity that the author had not been able to say something about the Tesla motor. The first form was hardly a practical alternating motor, because it required three leads; and if a third lead was run, a direct current might as well be used, with motors which it was known could work efficiently. But he thought that he had heard something about a lagging transformer, and he should like to know more. The author had called the Mordey alternator an iron-clad machine. He considered the word iron-clad a bad one to use. It was employed some time ago to denote a machine in which the iron was wrapped round the copper; the idea was that a line integral of the magnetic force of more than $0.4\,\pi$ time the ampere turns could somehow be obtained; but that fallacy had been dropped, and it was a pity to use the word for the Mordey machine, where the only object was to get the whole field through one exciting coil. The fault of the Kennedy machine, as far as it had any, was the want of ventilation. One of the great beauties of the

Mordey machine was that it could be run with a high-current density without fear. It had occurred to him that if an iron core were put into the Mordey machine, and the field-poles were arranged to come alternately, as in the Kennedy machine, instead of opposite, the power of the machine would be doubled without greatly increasing the expense. He agreed with Professor Forbes as to the credit due to Messrs. Gaulard and Gibbs, which, he thought, ought to be frankly acknowledged.

The Hon. Charles Parsons said, as the calculations in the paper were based on the sine-function, he would refer to a few figures in connection with a bipolar alternator, in which the sine-function was nearly realized. With two poles, and one continuous coil, or two half-coils, as on the drum of the armature, Fig. 8, the electromotive force was a sine-function, and the self-induction was considerable. In a contemplated machine, the output of which was intended to be 75,000 watts, the current to excite the magnets would be about 1 per cent. of the output, and the loss in resistance of the armature about 1½ per cent. This machine (Fig. 31) was not yet completed, but one of 30 units had been made and carefully tested. With regard to the question of hysteresis, he had made experiments upon the loss in the armature. A cylindrical armature-core, 5 inches in diameter, composed of one hundred iron plates to the circle, insulated from each other by thin paper, but not insulated from the spindle, was rotated in a bi-polar field. The energy required to rotate the core when subjected to different intensities of field, and at different speeds of rotation, was measured by independent means. In one experiment the same core was rotated, first at 8,500 revolutions per minute, and, secondly, at 6,500, the strengths of field being maintained in the two cases inversely proportional to the speeds; or, in other words, the electromotive force, had the core been wound with conducting wire, would have been identical in the two cases. It was found that the power lost by hysteresis at the lower speed was approximately 25 per cent. greater than at the higher speed. Taking the energy absorbed per revolution in the two cases, it followed that the loss per reversal was approximately proportional to the square of intensity of field. Some other experiments had been made which went to show that the viscous hysteresis, or the viscous resistance of the iron to rapid

Fig. 31

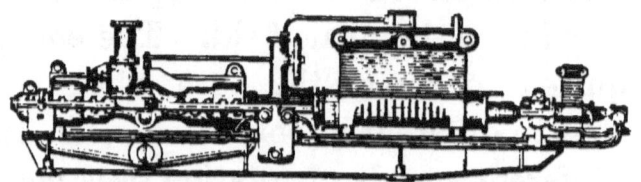

Turbo Electric-Generator for 75,000 watts for alternating currents.

change of polarity, was a comparatively small quantity. The intensity of field in the above experiments was between 5,000 and 7,000. At 6,000 revolutions of the apparatus the loss would be 2 HP., which was about 2 per cent. of the output, making a total electric efficiency of 95 or 96 per cent. for the dynamo. The machine was arranged with an electric-control governor, which was compounded so as to keep the electromotive force at the terminals constant at all loads. Though the machine revolved at a high rate of speed, the wear and tear was exceedingly small, the same bearings having been running for three years without any perceptible wear.

Mr. W. B. Esson observed that most of the points he might have referred to had been taken up by other speakers. The induction-curve in the Mordey alternator would differ considerably from the curves shown by Mr. Atkinson, for two or three reasons. The first was that the coil in the alternator was not merely a rectangle concentrated in one wire, as would be required to give the particular curve in 2, but a number of wires occupying a considerable angular width. Another reason was that, owing to the trapezoidal form of coil, the induction commenced very gradually, and rounded off the corners in the diagram. The peculiarity of that alternator was that the induction-curve of individual coils was not symmetrical, but as, in the machine, the coils were joined in pairs, it came out that, though considering any coil by itself there was a want of symmetry, when the coils were joined up all in series, the induction-curve of the whole machine was quite symmetrical; it was almost exactly a curve of sines. Mr. Mordey had measured the induction through a pair of coils for all different positions of the field-magnet in a whole period, and, as the curves showed that the induction lay within the curves of sines at some parts, so the curve for the electromotive force lay at some parts outside the curve of sines. He should like to have some definite information regarding the amount of power required to reverse the magnetism in the ring-cores of alternating dynamos for different degrees of magnetization. He had been making some investigations into the subject, and had found that to prevent the cores of alternate-current dynamos from over-heating, the thickness of the iron core for 100 periods per second should not exceed 9 centimeters, with a saturation of 7,000 C.G.S. units.

Mr. A. A. Campbell Swinton said there seemed to be an impression amongst engineers generally, that very high-speed machines like the Parsons must sacrifice durability. At Sir William Armstrong's works at Newcastle there were three Parsons machines, two of which ran at about 9,000 revolutions per minute, and the other at about 17,000. They had been running for three years, and neither the bearings nor the spindles had appreciably worn away. He thought that the brushes and the commutators were the only parts which had been renewed. Of course the Parsons machine was a special machine, and the bearings were specially designed to suit the high speed. The fact that high speed had not received much attention was not the fault of electricians, but of engineers. If engineers had produced engines that would run at these high speeds, no doubt electricians would have produced dynamos for them.

Mr. F. H. Nalder observed that the insulation resistance between the primary and secondary coils was important in the transformer. It was also of very great moment to apparatus in houses, which, if the insulation should break down, would be liable to cause fire. That in a measure was looked after by most makers, who provided covers for the transformers. Engineers should come to the same conclusion with respect to the minimum insulation resistance between primary and secondary coils, and also between the primary coils and the frame.

Professor W. E. Ayrton remarked that although some points in the paper might be not quite right, it was of a highly suggestive character. He wished to follow those who had preceded him in condemning the expression "mean electromotive force" employed in the paper. The expression had now a perfectly definite signification in connection with alternate-current circuits; it was, therefore, a mistake to use it as a substitute for another expression <u>which</u> had also a clear significance—"the square root of the mean square." The two things differ by many per cent., depending on what function the electromotive force was of the time. The mean electromotive force of an alternate-current and a direct-current dynamo similarly constructed, must be identical. Whether the wire was piled up all at one place, or whether spread uniformly round the armature, as in a direct-current dynamo, the mean electromotive force must necessarily be the same. The author spoke of "English lines of force." There was no such thing as an English line of force; lines of force were essentially cosmopolitan. One of the great charms that the absolute system of electrical and magnetic units possessed was its perfect harmony, and therefore any attempt to introduce new units should be deprecated. Experiments made at the Central Institution, for finding the coefficient of self-induction of a transformer, showed that in no case was there no lag between the impressed electromotive force and the primary current. Measurements had also been made by a number of students to ascertain the lag between the primary and the secondary current. The secondary circuit was sending a current through a non-inductive resistance. For all speeds varying from about 1,600 semi-alternations, up to 16,000 semi-alternations per minute, there appeared to be a lag of about 90° above the lag which ought to be got, which would really correspond to the maximum lag possible. In fact, there seemed to be a considerable amount of self-induction in a transformer, even when the secondary circuit was employed to send a current through glow-lamps, or a non-inductive circuit. Without lag, the wave of the secondary current must necessarily be 90° behind the wave of the primary; but instead of that, the former was 180° behind the latter, so that there were 90° extra lag in a transformer. The same result had been observed by Mr. Blakesley later on, and referred to in a paper read before the

Physical Society.* He did not, therefore, think that the arrangement could be taken as one which would give a non-self-induction in the external circuit. The author had stated (p. 61), "the coefficient of self-induction was found to be $L = 0.955$ in C.G.S. measure, and 0.0955 in practical units." C.G.S., of course, meant centimeter-gram-seconds; what "practical" units were he did not know. If they were the ordinary units of self-induction, corresponding with the true ohm, the volt, the ampere, and so on—then 1 practical unit was equal to 10^9 C.G.S. units or centimeters. The number 0.955 given by the author should have been divided by 10^9 or one thousand million; or more strictly, in consequence of the ohm not being the true ohm, the practical unit of self-induction was $99,777 \times 10^4$ centimeters. Hence the coefficient of self-induction referred to by the author was 0.9571×10^{-9} centimeters in practical units, which would be about

---

*Proceedings of the Physical Society of London, 1888, vol. ix. p. 287.

6,200 miles. Two years ago Professor Perry and he had suggested a name for the practical unit of self-induction, it not having previously received any name. If the practical unit were exactly a thousand million centimeters, the name quadrant would be an appropriate one, but it was about $\frac{1}{5}$ per cent. less than the earth's quadrant, and they had therefore suggested the name "secohm" (a contraction of second and ohm). Professor Ayrton next referred to a standard of self-induction which he exhibited on behalf of Professor Perry and himself, and which he thought was the first commercial standard of self-induction ever issued. It consisted of two coils of platinoid wire, one fixed in position, and the other pivoted so that it could be turned about a diameter, and placed in different positions relatively to the fixed coil. By putting the coils in different positions the arrangement had different and definite coefficients of self-induction, the values of which were indicated on the scale in secohms from 0.013 to 0.036

secohm. There was also an apparatus on the table familiar to some members, called a secohmmeter, for comparing two self-inductions. It rapidly alternated the battery connections and the galvanometer connections of a Wheatstone's bridge, and it rendered the measurement of self-induction, which before two years ago usually had been very unsensitive, having had to be effected with single impulses, as sensitive as the ordinary measurement of resistance with a Wheatstone's bridge. With a combination of those two instruments the coefficient of self-induction could at once be measured with accuracy, without any other apparatus except the ordinary Wheatstone's bridge. Fig. 32 showed symbolically the arrangement for measuring a coefficient of self-induction L_2. G C and B C were the galvanometer and battery commutators of the secohmmeter, which were rotated by turning the handle H. Balance for steady currents having been first obtained, the secohmmeter handle was rotated at any convenient speed, and the self-induction L_1 of the standard adjusted until balance was again obtained, when $\dfrac{L_2}{L_1} = \dfrac{r_2}{r_1}$. The continuous lines showed

Fig. 32

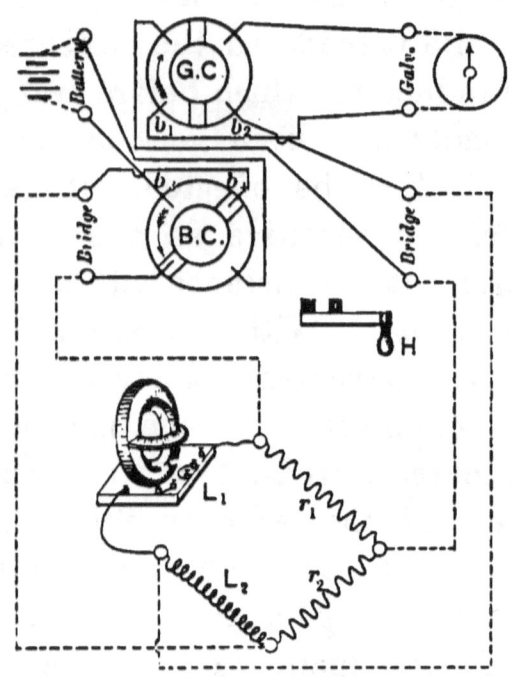

MEASUREMENT OF SELF-INDUCTION WITH SECOHM-
METER AND SECOHM STANDARD.

permanent connections in the secohm-meter, and the dotted lines wires employed to join its terminals with the Wheatstone's bridge.

Mr. T. H. Blakesley said that the paper had considerable breadth; but, like a dangerous piece of water, it was unequally deep. The subject was a very large one, and perhaps the author had not been able entirely to do justice to it, even in his comparatively long paper. Mr. Blakesley had himself done a good deal of work in diagrams, and the first remark he would make was that in Figs. 7, 8, and 9, the direction was continually reversed. In Fig. 7 the motion of the diagrams was supposed to be clockwise, in Fig. 8 it was the reverse; and in Fig. 9 it was supposed to be the same as in Fig. 7. The advantage of a diagram of that kind was that it not only represented the magnitude of certain quantities of electromotive force in general, but the relative positions of their maximum value. The author constantly called O A the current. It was true he guarded himself by saying that it was proportional to the current; but those who read the paper might suppose that he thought O A to be in phase with the current. It was not so. O R was the real effective electromotive force, requiring mere division by the resistance to represent the current, as in ordinary steady flow. In a good diagram, it only

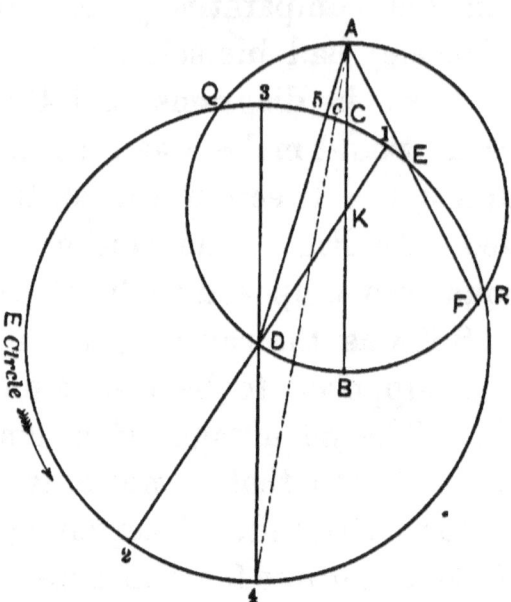

Fig. 33

needed a pair of dividers to work out results which, so far as the scale allowed, were as complete as the results of analysis. That was why they were so extremely valuable. He would now mention the details of the diagram (Fig. 33) with a view of fulfilling the characteristics which he had mentioned, and which represented all cases of the following problem:—Given two alternate-current machines coupled in series, and having given electromotive forces, and the same known period of alternation, a given resistance in the whole series and a given coefficient of self-induction for the whole series: how will any time-interval between the corresponding phases of the two machines (phase angle), affect (1) The work to be put into the system; (2) The work to be got out of the system; (3) The stability of the motion; (4) Efficiency (or the relation of (2) to (1)); (5) the work wasted in heating the circuit? Proceed thus:—With a pair of dividers lay down to any convenient scale, A B, proportional to the electromotive force of the first machine, and B C (in the same straight line but in the opposite direction), proportional to that of the second machine, and describe a circle upon A B as diameter. Adopting as the positive direction of rotation that which was counter-clockwise, set off the angle B A D in the negative direction, equal to that of electric lag (Tan B A D = $\dfrac{L\,\pi}{T\,R}$, where L = the coefficient of self-induction, R = the resistance of the circuit, and T = the semi-period). From B and C drop perpendiculars upon A D, cutting it in D and 5 (not given in Fig. 33 to avoid confusion of lines). D would obviously be on the circle already drawn. With D as center, and D 5 as radius, describe a circle, called the E-circle, cutting the former circle in Q and R. Then all possible phase-differences between the two machines might be represented by some angle A D E, where E was always upon the E-circle. For any point E the electromotive forces of the two machines were related in phase and magnitude, exactly as were A D, D E. For example, if E was at the point 5, the phase-angle was 180°; or the electromotive force of one machine was throughout its variations always opposed by the electromotive force of the other. Further, if for any point E, A E was joined and produced, if necessary, to cut in F the circle on A B as diameter, the values of A E, E F, A F, could be taken with the dividers, and measured on the scale first employed. These lines had the following properties:—

If R was the resistance of the circuit, then—

The power employed on the first machine (input) $= \dfrac{A E, A F}{2 R}$

The power heating the circuit - - - - - - $= \dfrac{A E^2}{2 R}$.

The power $\begin{Bmatrix}\text{given out by}\\ \text{employed on}\end{Bmatrix}$ $= \dfrac{A E, E F}{2 R}$. the second machine - -

The latter depending on whether E lay within or without the circle upon A B. If E lay within the circle upon A B, work would be given out by the second machine; that was, it might be employed as a motor. If, on the other hand, E lay upon that portion of the E-circle without the circle upon A B, power must be employed on the second machine from outside, and it would do part of the work of heating the circuit. Thus the two problems, of transmission of power and working in series, were really parts of the same general problem. The points Q and R were the positions of E where these problems passed one into the other, namely, where the output was zero. Through D draw a diameter to the E circle parallel to A B, or 3 D 4, and a second diameter through K, the bisection of A B, or 2 D K 1. It would be seen that these two diameters were equally inclined to A D. The points upon the E-circle so obtained had the following peculiarities:—If E coincided with 1, the maximum power would be given out by the second machine; if E coincided with 2, the maximum power existed which could be employed upon the second machine; if E coincided with 3, the minimum power existed of the first machine; if E coincided with 4, the maximum power existed of the first machine. Further, if E lay upon the semi-circle 1 Q 2, there was stability for the second machine; and if **E** lay upon the semi-circle 2 R 1, there was instability for the second machine; if **E** lay upon the semi-circle 3 2 4, there was instability for the first machine; and if **E** lay upon the semi-circle 4, 1, 3, there was stability for the first machine. Thus the region 1, 5, 3, had a double guarantee of stability. The angle 1 D 3 was twice that of electric lag; which showed one of the advantages of self-induction. Again, if a line from **A** to 4 cut the **E**-circle in 6, this would be the point for **E** giving the maximum efficiency in the transmission of power. It must lie in the region of stability. The point 5 had the property that the efficiency there had the ratio of the two electromotive forces, as in the case of steady currents, but it was not as great at 5 as at 6. Fig. 34 had been constructed in the same way as Fig. 33, and with the same electromotive forces, but with a larger angle of electric lag B A D. It was intended to show by an example that the point 3 did not necessarily lie within the circle upon A B. If circumstances were as in Fig. 34, then E might come between 3 and Q, and be outside the circle upon A B. Thus there would be, for such a point, stability for both machines at the same time that both

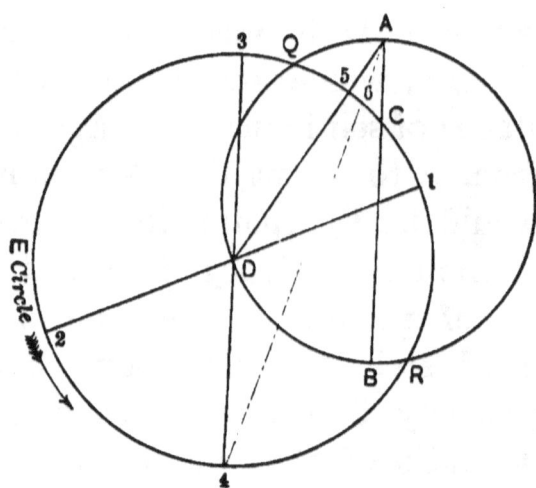

Fig. 34

machines would take part in heating the circuit. The possibility of such an arrangement had been denied by Dr. Hopkinson in this Institution, and wide currency had been given to his statement. It should be distinctly understood that the statement was not one of general truth, and with short periods and large self-induction, two machines in series might easily be run to work with stability. He had only been able to indicate the use of such diagrams as he recommended. A very slight addition to the diagrams enabled the effect of hysteresis to be taken into account. Hysteresis was the work absorbed in the magnetized masses in the field, upon change taking place in the degree or direction of magnetization. This was now a recognized fact, and well-known measurements of the effect had been made by Dr. Ewing and others. On the subject of transformers, he might mention a system of transformation which might be usefully employed in many practical ways. Suppose that an alternate-current machine (Fig. 35) played into a conductor possessing, besides resistance, various causes of self-induction. And suppose that at two points A, B, of the circuit were coupled the poles of a condenser. Then in general it would be found that: (1) The relative values of the current, in the sections near the machine and remote from it, became changed from unity, which was, of course, the relation before the introduction of the condenser: (2) There arose a difference in phase in the two portions of the circuit. Both these changes depended only upon: (i) the coefficient of self-induction in the remote section only; (ii) the resistance in the remote section only;

Fig. 35

(iii) the capacity of the condenser; (iv) the period; and not upon the coefficient of self-induction, or the resistance of the section near the generator, or of the generator itself. Under a proper arrangement of the four properties quoted, a considerable excess of current could be easily effected in the remote section over that in the nearer section. Mr. Nalder had been kind enough to give him the opportunity to test the results of the theory in practice, and he supposed they were the first people to see, so far as it was possible by means of a dynamometer to do so, a current in one conductor at some distance from the generating machine very much in excess of its value at points near to and in the machine itself. Not merely could such a transformation in relative magnitude be effected, but the phase relation could be adjusted, thus giving the method very wide applicability. For instance, the two portions of the circuit could have their phases thrown into quadrature with each other; while the values might be kept equal, which was the condition of things desirable in the Tesla motor, now obtained, as Professor Ayrton had said, by two circuits.

Captain P. Cardew, R. E., said that in the paper the reasoning was based upon the usual assumption that the varying quantities, magnetic induction, electromotive force, and current, followed the sinusoidal curve. That was admittedly only an approximation for the purpose of facilitating the calculations. It was probably nearly true in the case of machines with no iron in the generative portion, or armature, in which a copper conductor was dragged through alternating magnetic fields; but, in the case of machines in which the armature wires

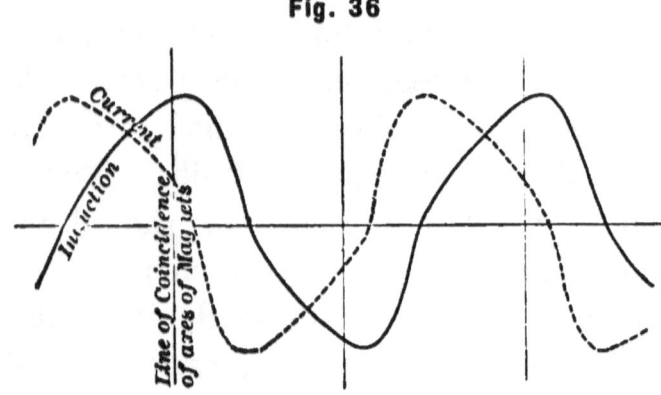

Fig. 36

were coiled on iron cores, the electromotive force being due to the variation of induction in those cores, he did not think that law of variation could hold. The maximum induction in the moving iron was not reached until after the point of nearest approach to the fixed magnet had been passed. When once the lines of force had begun to clear out of the iron, the action would go on rapidly, in spite of the opposing action of the induced current, owing to the molecules of iron assisting the demagnetization, the curve of induction being something like the firm line, while the curves of current were like the dotted line in Fig. 36. In the same way, if the moving iron was whirled past alternating polarities, the curves of induction and of current would still be unsymmetrical and unsymmetrically disposed with regard to the times of coincidence of the axes of the fixed and moving magnets (Fig. 37).

Mr. W. M. Mordey was glad the author had adopted his term "Alternator." The old term "alternating-current dynamo-electric machine" was very cumbrous. Alternators might be divided into two classes, those which had iron in the armatures, and those which had not. As his own machine, made by the Brush Corporation, and that of Ferranti, were the only examples of the latter class described by the author, he might be permitted to mention the reasons which had led him to avoid the use of iron. He thought that makers of dynamos would agree that almost all the ills that the dynamo was heir to were due to the presence of the iron in the armature. Its use in direct-current machines might be said to be a

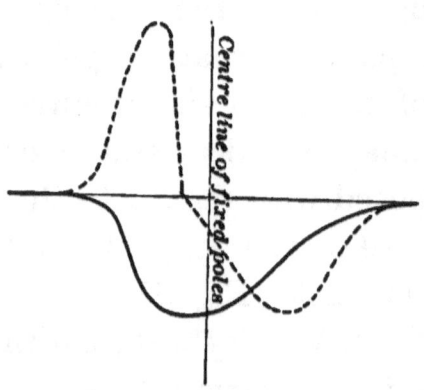

Fig. 37

necessary evil, as, although attempts had been made to do without it, all successful types of such machines required it for structural purposes. Iron in armatures when worked at a high magnetic density, and with rapid reversals or variations of magnetisms, became heated, and wasted a good deal of power. In alternators this objection applied with very much greater force than in direct-current machines, for in the latter the reversals of magnetism were comparatively slow. Thus the first result arrived at, in quite recent practice, was that the magnetic density that could be used in iron-cored alternators must only be about one-half that employed in direct current armatures. This alone meant a considerable increase in the size of armatures, without any gain in output or efficiency. Although the loss per cubic inch was reduced by decreasing the magnetic density, the armature had to be made larger to compensate for it, and the total loss was actually increased, not reduced. Iron was used to reduce the magnetic resistance, to afford mechanical support, and to introduce self-induction into the circuit. The latter, an evil in itself, was said to be a modern necessity, caused by the convenience of working alternators in parallel. If iron was indispensable for this purpose, which he was not prepared to admit, it could readily be inserted in some part of the circuit where ample space and cooling surface could be provided, and from which it could be easily removed when not required, that was when it was only necessary to run one machine. The armature was the very worst place for iron. He ventured to think it was much the best to make the armature stationary, as he had done. It then had only to resist the tangential drag of the field. He thought a great mistake was made in some alternators in using Pacinotti projections. In all cases there should be, as nearly as possible, a steady magnetic flux in the field. This could not be done if projections were used. The Zipernowsky and Parsons machines were faulty in this respect. Much had been said about the form of the wave yielded by alternators. He had some time ago made an experiment with the first of his machines, using the method which was now described by Professor Ayrton, and found that the curve had the form shown by Fig. 38, which was almost a sine-curve. On the subject of the lag in transformers, to which Professor Ayrton had also alluded, he might be allowed to mention that, in the discussion of another paper by the author, he had first stated the fact,* and had described a simple experiment, showing that the primary and secondary currents reached their maxima and minima practically at the same time.

Sir James N. Douglass considered that the discussion had thrown a considerable amount of light on the relative advantages of direct and alternate-current machines for lighting and power purposes. In 1862 the French lighthouse authorities adopted the alternate-current machine of the Alliance Company at the two lighthouses of Cape la Hève; and in 1866, Holmes produced his alternate-current machine, and a pair of these were made for the Trinity House and exhibited at the Paris International Exhibition of 1867. These machines were installed at the Souter Point Lighthouse in 1870, where they had worked regularly and

---

* Journal of the Institution of Electrical Engineers, vol. xvii., p. 215.

efficiently ever since, and without a single failure or the necessity for any repair. In 1877, the Trinity House made a series of competitive trials at the South Foreland, with the Holmes and Alliance alternating, and the more powerful direct-cur-

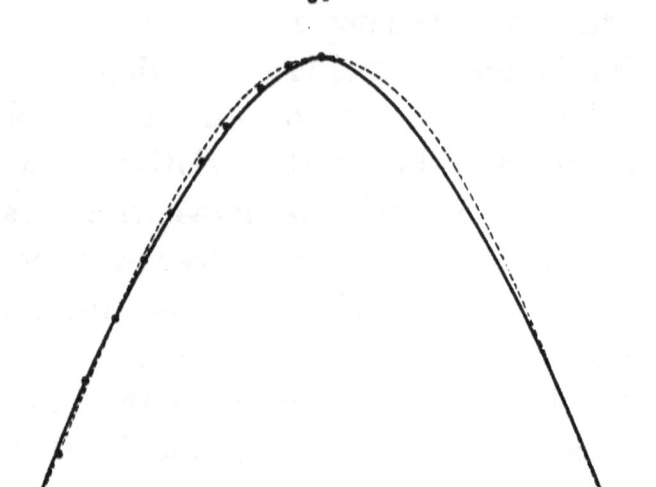

Fig. 38

MORDEY-VICTORIA ALTERNATOR. OPEN CIRCUIT CURVE OF INDUCTION, HALF PERIOD. THE DOTTED LINE IS A SINE-CURVE.

rent machines of Gramme and Siemens, which resulted in the adoption of the Siemens machines for the two Lizard lighthouses in 1878. In 1879, the alternate-current machine of Baron de Meritens was tested at the Royal Institution by Dr. Tyndall and himself, and one was purchased by the Trinity House in 1880, and sent to the Lizard lighthouses, where it had been in successful operation every night since that date, giving no more trouble than the original alternate-current machines of the Alliance Company and Holmes. Three of these machines of larger size, giving currents of about 160 amperes and 42 volts, were purchased by the Trinity House, and used at the trials, at the South Foreland, in 1884-85, into the relative merits of electricity, gas, and mineral oil as lighthouse illuminants. These machines gave great satisfaction, and two of them had since been permanently installed at St. Catharine's lighthouse.

Mr. G. Kapp, in reply, said he had first thought that his paper was more scientific than it should have been, but the discussion had beaten him completely in the amount of science which had been developed. Considerable importance seemed to have been attached to the nomenclature in the paper. Whether a certain electromotive force should be called "mean," "average," "equivalent," the "square root of the mean squares," or by any other name, might be a fit subject for discussion from a pure academic point of view, but practical engineers were satisfied with the terms ordinarily in use, provided definitions were added to prevent misunderstanding. The diagrams had not been designed with a view of finding out everything which could possibly be ascertained about alternators, but simply as mental tools, perfected only so far as necessary for the work of designing alternators, but not further. The statement in the paper, that there was no self-induction in a bank of transformers working glow-lamps, although described as incorrect by Professor Ayrton, had in reality been confirmed by the remarks that gentleman had made, and also by what fell from Mr. Mordey. Both had said virtually the same thing, namely, that the lag was nearly 180° between the phases of primary and secondary current.

But the lag between the induced counter electromotive force in the primary and the secondary current was also 180°, and therefore the phases of terminal electromotive force, induced electromotive force, and current in the primary, must very nearly coincide, which was only possible if there was practically no self-induction. He wished to offer an explanation with regard to the units used in the paper. His reason for adopting these units was that they were convenient, and since he first brought them forward many others had adopted them. All knew that the scientifically correct unit was inconvenient for the shop, and as the object of practical engineers was to build machines in the easiest way, he had adopted that system of units which many years' practice had shown him to be the most convenient. Professor Forbes had objected to the parallel working of dynamos, chiefly on the ground that it was necessary to cut out certain districts in case of fire. Mains might be joined in parallel at the stations, but they need not be joined outside where the fire might occur. There was no objection to working all the circuits from a pair of omnibus bars in the station, but they should be disconnected in the districts to be lighted. Another objection urged was the drop of external pressure with an increase of current, provided the exciting current was not varied. That was a serious objection when there was one dynamo machine only at work, and when a large part of the total load would, on certain occasions, be taken off suddenly. For example, in ship-lighting it often happened that the lamps were divided between a few circuits, and a large number of lamps must thus be switched off at one instant. This would, with an alternator having a large self-induction, produce in the remaining lamps a great jump. But these machines were not intended for ship-lighting or small isolated installations; they were intended for large central stations where there were many thousand lamps, and where the effect of switching on or off even as many as one hundred lamps simultaneously would be infinitesimal in comparison with the large number of lamps alight. He did not think that a switch that contained mercury troughs which would bob up and down was as good as a switch with solid contacts. The switches used at the Grosvenor Gallery were of the latter type, and in his opinion far superior to the mercury switches which Professor Forbes had praised so much. Professor Forbes was a little too severe on English engineers when he said that they merely copied foreign dynamos. It was quite true that in a certain sense all machines resembled each other, in so far namely as all contained field-magnets and an armature; but this was also the case with continuous-current dynamos, of which many different types were nevertheless recognized. He did not think high frequency was objectionable on the ground of parallel working, but it was on the grounds which he had mentioned in the paper. He was sorry Professor Forbes had found fault with the transformers made in England. He appeared to have a low opinion of English engineers since he had seen some of the central stations on the Continent. His contention was that in England electricians were employing too long a magnetic circuit. But all these transformers which Professor Forbes criticised adversely were not made yesterday; they were begun to be made three years ago. Many hundreds were at work, and they had been altered and altered till the present forms had been developed. The magnetic circuit was long, but the electric circuit was short, and a short electric circuit was far more important than a short magnetic circuit. As a matter of fact, even the Zipernowsky transformer, which Professor Forbes admired so much, had a long magnetic circuit. Three or four years ago, at the Inventions Exhibition, that wonderful transformer with a specially short magnetic circuit was first seen. It was a Gramme ring turned inside out. It was stated to be a marvelous improvement; but not one of those transformers was now at work; all had been exchanged for the ordinary Gramme ring, with iron inside and copper outside, because in this form the electric circuit was short; consequently, the regulating power of the transformer was better than it had been previously. The other question raised about transformers was whether a high or a low frequency was advantageous. A fairly high frequency was advisable; but above a certain limit, say 70 complete periods per second, an increase of frequency did not reduce the size of the transformer. On the table there was one of his transformers, weighing 200 lbs., and feeding fifty lamps, giving a weight of 4 lbs. per lamp. It was built for a frequency of 70. The Westinghouse transformer (taking the figures from a paper by Professor Forbes at the last meeting of the British Association) was for forty lamps, and it weighed 160 lbs., or 4 lbs. per lamp. In that case the frequency was 133 as against his 70. That was a proof that the increase of frequency did not diminish the weight of the transformers. Professor Forbes had brought a diagram showing the output from the Grosvenor Station (Fig. 28), and he argued that the transformers by themselves absorbed an energy equal to four thousand lamps. That was surely not to be taken as a serious argument. The diagram was not one of HP., but one of current sent out. Now the current sent out of course depended on the lamps which were alight. The argument was that in the early hours of the morning there were no lamps on circuit. But that was not so certain. In the case of stations supplying electric light by contract, for a fixed rental without a meter, people were wasteful with the lamps and allowed them to burn all night long. Professor Forbes had overlooked one important circumstance, namely, that if the transformers were on the circuit without giving off current to the lamps, there was a lag of 70°, 80° or more between the terminal electromotive force and the primary current in each transformer. The energy must therefore not be taken as represented by the product of electromotive force and current, as given in the diagram, but by this product multiplied by the cosine of the angle of lag. On purely theoretical grounds all the transformers now in use were equally good and their efficiency was high. Professor Ayrton had tested some, and found that it was as high as 95 per cent. The difference in the transformers was in their details, such as would secure low first cost, facility of manufacture, ease of changing a coil, freedom from humming, ability to keep cool, and, above all things, absolute safety of insulation.

CORRESPONDENCE.

Mr. J. D. F. Andrews, in reference to the author's observations on eddy currents (p. 64), and his inability "to suggest an entirely satisfactory explanation for the effect of the iron core in reducing eddy currents in the copper," stated that he had contributed the results of a series of experiments in a letter to Industries,* in which he showed that the strength or amount of eddy currents depended on the speed with which the wire moved into and out of the magnetic field, and that they could be greatly reduced by shaping the magnet-poles so that the wire gradually entered and left the field. The shape of the conductor, and its position relative to the armature core and the field-magnets, had a greater bearing on the reduction of eddy currents than the shape of the field-magnets; and round wire of any diameter was nearly proof against eddy currents if the machine was otherwise correctly proportioned.

Mr. A. Du Bois-Reymond remarked that some minor points in the diagrams seemed open to criticism. With regard to Figs. 9 and 10, he could not fully comprehend the author's meaning when he said, " the length O A, that is to say,

---

* Vol. iv. 1888, p. 9.

the current," and accordingly defined the mechanical output of the machine as "the product O A and O D." As far as he could make out, O A was the resultant electromotive force obtained by the addition of the terminal and induced electromotive forces E_t and E, and by the author's own showing (Fig. 7) the current ought to be the resultant of this and of the electromotive force due to self-induction, that was, it ought to coincide with O R; or assuming the terminal resistance to be one, it would be equal to O R. Accordingly, the mechanical work generated would be the product of O R and O D, namely, E. I. cos B O F, I standing for the current. Moreover, he should prefer assuming some definite position of E_t to start from as being more conformable to reality, aud letting E shift about, thereby altering the positions and magnitudes of the other quantities. There would then be perfect agreement between the author's diagram and the geometrical solution of the same problem explained by him in a paper, "On the difficulties in the way of transmitting power by alternating currents."* The use to which the author put the diagrams, namely, to determine the safety of running a given machine in parallel with others, was admirable, and quite put an end to the doubts hitherto entertained by practitioners on this head.

Professor J. A. Ewing said that at p. 65 of the paper there was a remark about the dissipation of energy in the armatures of dynamos through magnetic hysteresis, from which it seemed to him that the author had perhaps somewhat misunderstood in one respect the position he had taken up in the paper there referred to. What he had shown was that the energy dissipated in iron, through magnetic hysteresis in reversals of magnetism, might be very much reduced if the metal were subjected to vibration during the process. When a soft iron wire, for example, was sharply

---

* The Telegraphic Journal and Electrical Review vol. xxiv. 1889, p. 112; and Elektrotechnische Zeitschrift vol. x. 1889, p. 1.

tapped during repeated reversals of its magnetism, there was practically no dissipation of energy through hysteresis, and even a very slight mechanical disturbance had some influence in reducing the dissipation. In applying this result to dynamos, all that he had said was this: "Hence, in a dynamo, where vibration occurs to a greater or less degree whenever the machine is running, the energy dissipated through changes of magnetization is even less than these experiments on still metal would lead us to expect."* He had meant, in these words, to draw attention to the fact that whatever vibration occurred in the armature of a dynamo would tend to lessen the dissipation of energy, produced by hysteresis, below the value measured by his experiments on still metal. He had not meant to suggest that any large effect of this kind was actually produced, and in fact he agreed with the author in believing that in a well-balanced dynamo the vibration was not likely to be seriously influen-

---

* Phil. Trans. Royal Society, 1885, p. 554.

tial, so that the values of the dissipation calculated from his experiments were probably not too high for application to such a case. With reference to "viscous hysteresis" (p. 67), he had made no direct observations showing that the energy dissipated through hysteresis increased with the speed at which the cyclic change of induction was performed. So far as any such observations had been made, he understood that the results had been negative, and that the heating effect per cycle had been sensibly the same whether the same cycle was repeated with greater or with less frequency. But there was a certain amount of evidence that magnetization took some time to be fully developed in iron after the magnetizing force was applied, especially when the magnetizing force was low, and so far as this was true it must have the effect of increasing the dissipation of energy in cycles of high frequency, since larger changes of magnetic force would then be required to produce equal changes of magnetic induction.

Dr. John Hopkinson said that one of the questions with which the author dealt was how many alternations per second was it most appropriate to use for distribution by transformers? It was clear that the answer depended upon a balance of advantages. No one would use exceedingly few alternations; no one would advocate a greater number than some superior limit much less than the greatest it was possible to produce. In favor of a high frequency was the fact that, for a given efficiency, the transformer would be cheaper to manufacture. On the other hand, in favor of a low frequency, the whole of the conductor was not equally used with alternating currents. The author had alluded to the question of a viscous hysteresis in magnetization of iron. Dr. Hopkinson doubted the propriety of the term viscous hysteresis, but he knew what was meant. There was, he believed, no satisfactory evidence of the existence of so-called viscous hysteresis, and it was certainly a phenomenon by no means so marked as true hysteresis when the magnetism of iron was reversed. In his lecture before the Institution in 1883, he not only explained why alternate-current machines could be run parallel but also mentioned that an alternate-current machine could be run as a motor. The theory of both cases had been fully developed by himself in a paper read before the Society of Telegraph Engineers;* and, at his suggestion, Professor Adams successfully tried the experiment of running a machine as a motor at the South Foreland. In 1887 he presented to the Royal Society two short papers.† The first related to transformers, and showed how to treat them, taking proper account of the true magnetic properties of the iron. The second treated the theory of alternate-current machines; showed how the true differential equation of currents could be

---

* Journal of the Society of Telegraph Engineers and Electricians, vol. xiii. 1884, p. 496.

† Proceedings of the Royal Society, vol. xlii. pp. 164 and 167.

obtained, and that it differed materially, in some cases, from the linear equation generally used, but was of a character not convenient for practical use.

Mr. Elihu Thomson, in regard to the relation between the width of the field-poles and the armature-winding, observed that the author's statement that "width of poles equal to half the pitch, smooth armature, and winding covering only one-half the surface," was most frequently met with in practice, was without doubt true. But was it the best practice? While designing such machinery, it occurred to him that a higher yield in electromotive force would be had from a given armature, particularly under full load, or "dynamic" working, as distinguished from open circuit, or "static" condition, if the open space in the interior of each coil was made less than the width of the pole-face. It was seen that this must result as a consequence of the crowding or distortion of the field-lines forward by the currents in the armature-coils, having the practical effect of somewhat narrowing the effective pole-face presented to the armature. A trial machine was built, on the type in use by the Thomson-Houston Electric Company, which consisted of a field structure with inward radial cores and a cylindrical armature with very flat coils laid on and bound down, their curved ends out of the field; not, however, being turned down toward the shaft, as is the case

Fig. 39

with the Westinghouse machine. The field-poles had a width approximately equal to one-half the pitch, the armature being smooth. In the trial machine the coils were laid on so as to practically cover the whole exterior of the armature-core, or they were wound with but a narrow line of open space in the center of each coil, as in Fig. 39. Arrangements were made that the inner turns might be progressively cut out, so as to virtually remove wire from each coil. Under static condition, or no load, the terminal electromotive force was practically unchanged, even though considerable fractions of wire-turns were cut out from the center of each coil. But under a load a limit was soon found, or a point reached, in which the electromotive force was the maximum under a given excitation of the field. This maximum was reached after a few turns had been removed or cut out; and the open space in the center of the coil so obtained was adopted for use in the apparatus subsequently manufactured on the large scale. The vacant space was less than one-half the width of the pole-face. The relations would be as shown in Fig. 40. He regarded this result as in a measure growing out of the condition of displacement of lines, indicated in Fig. 41, where the current in the armature-coil drew the lines together, leaving the field, or bunched them through its own center in the position indicated. In dealing with the heating of armature conductors by eddy-currents, the author had drawn a distinction between coils with and without iron cores, showing the need of an explanation of the differences noted. His own view of the matter had for some time been this: It is well known that, in a transformer with a well-closed iron circuit, very large sections of conductors might be employed

Fig. 40   Fig. 41

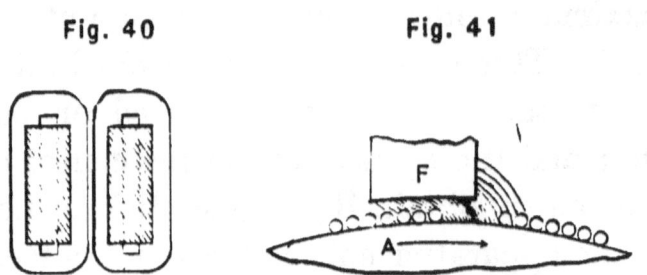

in the coils without introducing trouble from eddy-currents. The extreme case occurred in transformers for electric welding, in which the section of the single secondary turn might be many square inches without showing the effects of eddy-currents. The reason, as it appeared to him, for such a condition was that the development of lines, taking place as circular magnetism around the primary conductors, was carried out with great speed to the lodgment of the lines in the iron shell or core. Thus all parts of the secondary conductor were cut at almost the same instant, by each line evolved, in its passage to the iron core, the same actions being repeated at each alternation. The great speed at which the lines passed the conductor in such cases, in going to the iron core, left no time during which differences of electromotive force in parts of the conductor might act to induce eddy-currents. Now, when a coil without an iron core passed through a field, the speed of cutting did not exceed the speed of motion, so that if the lines being cut by a conductor were denser at one point than at another, eddy-currents resulted. Provide the coil with an iron core and the lines remain in the core until dragged out, as it were, by the movement given to it in front of the field-cores, at which time the lines move through the space between the successive cores with great speed, and cut all parts of the conductor at the

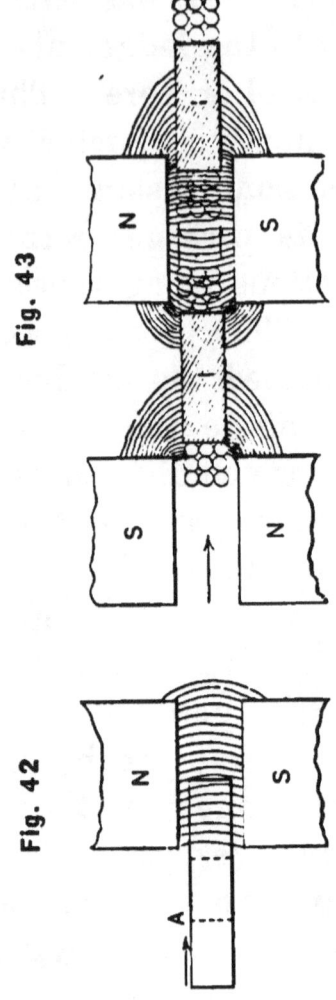

same instant. The lines snap out of the leaving core, as it were, and pass quickly to the incoming core, producing a weak field between the two composed of few lines moving at great speeds. Figs. 42 and 43 represented the two conditions comparatively. Fig. 42 showed a coil without core cutting lines between N and S; and Fig. 43 the progressive movement of coils with iron cores through the field. The rapid movement of the lines backward across the space between the iron armature-cores, or projections, I I, Fig. 43, as the armature moved forward, pointed also to the necessity for lamination of the field-poles when such cores or projections were used. The turns of armature-conductor nearest the enclosed iron core would be subject to eddy-currents, on account of the field density being made more irregular by the bending of the lines leaving the iron. He was of opinion that the use of "horns" on pole-pieces was not akin to the effects just mentioned, excepting in so far as the entering and leaving field was more spread out or graduated in density, and so placed the parts of a large section armature-conductor in substantially equal field densities at any instant during movement into or out of the field. This was not the case where no horns were used, or where the field was abrupt and permitted, say, one-half the section of the armature-conductor to be at any instant cutting dense lines, and the other half to be moving in a field of little density. Making the conductor of insulated strands, and twisting the strands, was the remedy indicated.

₊ *Any book in this Catalogue sent free by mail on receipt of price.*

VALUABLE
SCIENTIFIC BOOKS

PUBLISHED BY

D. VAN NOSTRAND COMPANY,

23 MURRAY STREET AND 27 WARREN STREET, N. Y.

ADAMS (J. W.) Sewers and Drains for Populous Districts. Embracing Rules and Formulas for the dimensions and construction of works of Sanitary Engineers. Second edition. 8vo, cloth..$2 50

ALEXANDER (J. H.) Universal Dictionary of Weights and Measures, Ancient and Modern, reduced to the standards of the United States of America. New edition, enlarged. 8vo, cloth.. 3 50

ATWOOD (GEO.) Practical Blow-Pipe Assaying. 12mo, cloth, illustrated.. 2 00

AUCHINCLOSS (W. S.) Link and Valve Motions Simplified. Illustrated with 37 wood-cuts and 21 lithographic plates, together with a Travel Scale and numerous useful tables. 8vo, cloth... 3 00

AXON (W. E. A.) The Mechanic's Friend: a Collection of Receipts and Practical Suggestions Relating to Aquaria—Bronzing—Cements—Drawing—Dyes—Electricity—Gilding—Glass-working — Glues — Horology - Lacquers —Locomotives—Magnetism—Metal-working - Modelling — Photography—Pyrotechny—Railways—Solders—Steam-Engine—Telegraphy—Taxidermy—Varnishes—Waterproofing, and Miscellaneous Tools, Instruments, Machines, and Processes connected with the Chemical and Mechanic Arts. With numerous diagrams and wood-cuts. Fancy cloth 1 50

BACON (F. W.) A Treatise on the Richards Steam-Engine Indicator, with directions for its use. By Charles T. Porter. Revised, with notes and large additions as developed by American practice; with an appendix containing useful formulæ and rules for engineers. Illustrated. Third edition. 12mo, cloth..................................... 1 00

D. VAN NOSTRAND'S PUBLICATIONS.

BARBA (J.) The Use of Steel for Constructive Purposes; Method of Working, Applying, and Testing Plates and Brass. With a Preface by A. L. Holley, C.E. 12mo, cloth. $1 50

BARNES (Lt. Com. J. S., U. S. N.) Submarine Warfare, offensive and defensive, including a discussion of the offensive Torpedo System, its effects upon Iron-Clad Ship Systems and influence upon future naval wars. With twenty lithographic plates and many wood-cuts. 8vo, cloth............ 5 00

BEILSTEIN (F.) An Introduction to Qualitative Chemical Analysis, translated by I. J. Osbun. 12mo, cloth.......... 75

BENET (Gen. S. V., U. S. A.) Electro-Ballistic Machines, and the Schultz Chronoscope. Illustrated. Second edition, 4to, cloth.. 3 00

BLAKE (W. P.) Report upon the Precious Metals: Being Statistical Notices of the principal Gold and Silver producing regions of the World, represented at the Paris Universal Exposition. 8vo, cloth.................................. 2 00

—— Ceramic Art. A Report on Pottery, Porcelain, Tiles, Terra Cotta, and Brick. 8vo, cloth...................... 2 00

BOW (R. H.) A Treatise on Bracing, with its application to Bridges and other Structures of Wood or Iron. 156 illustrations. 8vo, cloth.. 1 50

BOWSER (Prof. E. A.) An Elementary Treatise on Analytic Geometry, embracing Plane Geometry, and an Introduction to Geometry of three Dimensions. 12mo, cloth....... 1 75

—— An Elementary Treatise on the Differential and Integral Calculus. With numerous examples. 12mo, cloth......... 2 25

BURGH (N. P.) Modern Marine Engineering, applied to Paddle and Screw Propulsion. Consisting of 36 colored plates, 259 practical wood-cut illustrations, and 403 pages of descriptive matter, the whole being an exposition of the present practice of James Watt & Co., J. & G. Rennie, R. Napier & Sons, and other celebrated firms. Thick 4to vol., cloth 10 00
Half morocco...15 00

BURT (W. A.) Key to the Solar Compass, and Surveyor's Companion; comprising all the rules necessary for use in the field; also description of the Linear Surveys and Public Land System of the United States, Notes on the Barometer, Suggestions for an outfit for a survey of four months, etc. Second edition. Pocket-book form, tuck............. 2 50

BUTLER (Capt. J. S., U. S. A.) Projectiles and Rifled Cannon. A Critical Discussion of the Principal Systems of Rifling and Projectiles, with Practical Suggestions for their Improvement, as embraced in a Report to the Chief of Ordnance, U. S. A. 36 plates 4to, cloth...................... 6 00

D. VAN NOSTRAND'S PUBLICATIONS. 3

CAIN (Prof. WM.) A Practical Treatise on Voussoir and Solid
and Braced Arches. 16mo, cloth extra $1 75

CALDWELL (Prof. GEO. C.) and BRENEMAN (Prof. A. A.)
Manual of Introductory Chemical Practice, for the use of
Students in Colleges and Normal and High Schools. Third
edition, revised and corrected. 8vo, cloth, illustrated. New
and enlarged edition.. 1 50

CAMPIN (FRANCIS). On the Construction of Iron Roofs. 8vo,
with plates, cloth 2 00

CHAUVENET (Prof. W.) New method of correcting Lunar
Distances, and improved method of finding the error and
rate of a chronometer, by equal altitudes. 8vo, cloth...... 2 00

CHURCH (JOHN A.) Notes of a Metallurgical Journey in
Europe. 8vo, cloth... 2 00

CLARK (D. KINNEAR, C.E.) Fuel: Its Combustion and
Economy, consisting of Abridgments of Treatise on the
Combustion of Coal and the Prevention of Smoke, by C.
W. Williams; and the Economy of Fuel, by T S. Prideaux. With extensive additions on recent practice in the
Combustion and Economy of Fuel: Coal, Coke, Wood,
Peat, Petroleum, etc. 12mo, cloth... .... 1 50

—— A Manual of Rules, Tables, and Data for Mechanical
Engineers. Based on the most recent investigations. Illustrated with numerous diagrams. 1,012 pages. 8vo, cloth... 7 50
Half morocco... 10 00

CLARK (Lt. LEWIS, U. S. N) Theoretical Navigation and
Nautical Astronomy. Illustrated with 41 wood-cuts. 8vo,
cloth 1 50

CLARKE (T. C.) Description of the Iron Railway Bridge over
the Mississippi River at Quincy, Illinois. Illustrated with
21 lithographed plans. 4to, cloth 7 50

CLEVENGER (S. R.) A Treatise on the Method of Government Surveying, as prescribed by the U. S. Congress and
Commissioner of the General Land Office, with complete
Mathematical, Astronomical, and Practical Instructions for
the use of the United States Surveyors in the field. 16mo,
morocco 2 50

COFFIN (Prof J. H. C.) Navigation and Nautical Astronomy. Prepared for the use of the U. S. Naval Academy.
Sixth edition. 52 wood-cut illustrations. 12mo, cloth...... 3 50

COLBURN (ZERAH). The Gas-Works of London. 12mo,
boards..... 60

COLLINS (JAS. E.) The Private Book of Useful Alloys and
Memoranda for Goldsmiths, Jewellers, etc. 18mo, cloth... 50

CORNWALL (Prof. H. B.) Manual of Blow Pipe Analysis, Qualitative and Quantitative, with a Complete System of Descriptive Mineralogy. 8vo, cloth, with many illustrations. N. Y., 1882 .... $2 50

CRAIG (B. F.) Weights and Measures. An account of the Decimal System, with Tables of Conversion for Commercial and Scientific Uses. Square 32mo, limp cloth 50

CRAIG (Prof. THOS.) Elements of the Mathematical Theory of Fluid Motion. 16mo, cloth 1 25

DAVIS (C. B.) and RAE (F. B.) Hand-Book of Electrical Diagrams and Connections. Illustrated with 32 full-page illustrations. Second edition. Oblong 8vo, cloth extra 2 00

DIEDRICH (JOHN). The Theory of Strains: a Compendium for the Calculation and Construction of Bridges, Roofs, and Cranes. Illustrated by numerous plates and diagrams. 8vo, cloth 5 00

DIXON (D. B.) The Machinist's and Steam-Engineer's Practical Calculator. A Compilation of useful Rules, and Problems Arithmetically Solved, together with General Information applicable to Shop-Tools, Mill-Gearing, Pulleys and Shafts, Steam-Boilers and Engines. Embracing Valuable Tables, and Instruction in Screw-cutting, Valve and Link Motion, etc. 16mo, full morocco, pocket form ...(In press)

DODD (GEO.) Dictionary of Manufactures, Mining, Machinery, and the Industrial Arts. 12mo, cloth 1 50

DOUGLASS (Prof S. H.) and PRESCOTT (Prof. A. B.) Qualitative Chemical Analysis. A Guide in the Practical Study of Chemistry, and in the Work of Analysis. Third edition. 8vo, cloth 3 50

DUBOIS (A. J.) The New Method of Graphical Statics. With 60 illustrations. 8vo, cloth 1 50

EASSIE (P. B.) Wood and its Uses. A Hand-Book for the use of Contractors, Builders, Architects, Engineers, and Timber Merchants. Upwards of 250 illustrations. 8vo, cloth. 1 50

EDDY (Prof. H. T.) Researches in Graphical Statics, embracing New Constructions in Graphical Statics, a New General Method in Graphical Statics, and the Theory of Internal Stress in Graphical Statics. 8vo, cloth 1 50

ELIOT (Prof. C. W.) and STORER (Prof. F. H.) A Compendious Manual of Qualitative Chemical Analysis. Revised with the co-operation of the authors. By Prof. William R. Nichols. Illustrated. 12mo, cloth 1 50

ELLIOT (Maj. GEO. H., U. S. E.) European Light-House Systems. Being a Report of a Tour of Inspection made in 1873. 51 engravings and 21 wood-cuts. 8vo, cloth 5 00

D. VAN NOSTRAND'S PUBLICATIONS.

ENGINEERING FACTS AND FIGURES. An Annual Register of Progress in Mechanical Engineering and Construction for the years 1863-64-65-66-67-68. Fully illustrated. 6 vols. 18mo, cloth (each volume sold separately), per vol...$2 50

FANNING (J. T.) A Practical Treatise of Water-Supply Engineering. Relating to the Hydrology, Hydrodynamics, and Practical Construction of Water-Works in North America. Third edition. With numerous tables and 180 illustrations. 650 pages. 8vo, cloth.................................... 5 00

FISKE (BRADLEY A., U. S. N.) Electricity in Theory and Practice. 8vo, cloth.. 2 50

FOSTER (Gen. J. G., U. S. A.) Submarine Blasting in Boston Harbor, Massachusetts. Removal of Tower and Corwin Rocks. Illustrated with seven plates. 4to, cloth......... 3 50

FOYE (Prof. J. C.) Chemical Problems. With brief Statements of the Principles involved. Second edition, revised and enlarged. 16mo, boards.. 50

FRANCIS (JAS. B., C E.) Lowell Hydraulic Experiments: Being a selection from Experiments on Hydraulic Motors, on the Flow of Water over Weirs, in Open Canals of Uniform Rectangular Section, and through submerged Orifices and diverging Tubes. Made at Lowell, Massachusetts. Fourth edition, revised and enlarged, with many new experiments, and illustrated with twenty-three copperplate engravings. 4to, cloth...................................... 15 00

FREE-HAND DRAWING. A Guide to Ornamental Figure and Landscape Drawing. By an Art Studen.. 18mo, boards... 50

GILLMORE (Gen. Q. A.) Treatise on Limes, Hydraulic Cements, and Mortars. Papers on Practical Engineering, U. S. Engineer Department, No. 9, containing Reports of numerous Experiments conducted in New York City during the years 1858 to 1861, inclusive. With numerous illustrations. 8vo, cloth.. 4 00

—— Practical Treatise on the Construction of Roads, Streets, and Pavements. With 70 illustrations. 12mo, cloth....... 2 00

—— Report on Strength of the Building Stones in the United States, etc. 8vo, illustrated, cloth 1 50

—— Coignet Beton and other Artificial Stone. 9 plates, views, etc. 8vo, cloth.. 2 50

GOODEVE (T. M.) A Text-Book on the Steam-Engine. 143 illustrations. 12mo, cloth............................... 2 00

GORDON (J. E. H.) Four Lectures on Static Induction. 12mo, cloth.. 80

GRUNER (M. L.) The Manufacture of Steel. Translated from the French, by Lenox Smith, with an appendix on the Bessemer process in the United States, by the translator. Illustrated. 8vo, cloth $3 50

HALF-HOURS WITH MODERN SCIENTISTS. Lectures and Essays. By Professors Huxley, Barker, Stirling, Cope, Tyndall, Wallace, Roscoe, Huggins, Lockyer, Young, Mayer, and Reed. Being the University Series bound up. With a general introduction by Noah Porter, President of Yale College. 2 vols 12mo, cloth, illustrated 2 50

HAMILTON (W. G.) Useful Information for Railway Men. Sixth edition, revised and enlarged 562 pages, pocket form. Morocco, gilt.. 2 00

HARRISON (W. B.) The Mechanic's Tool Book, with Practical Rules and Suggestions for Use of Machinists, Iron-Workers, and others. Illustrated with 44 engravings. 12mo, cloth.. 1 50

HASKINS (C. H.) The Galvanometer and its Uses. A Manual for Electricians and Students. Second edition. 12mo, morocco... 1 50

HENRICI (OLAUS). Skeleton Structures, especially in their application to the Building of Steel and Iron Bridges. With folding plates and diagrams. 8vo, cloth.................... 1 50

HEWSON (WM.) Principles and Practice of Embanking Lands from River Floods, as applied to the Levees of the Mississippi. 8vo, cloth...................................... 2 00

HOLLEY (ALEX. L.) A Treatise on Ordnance and Armor, embracing descriptions, discussions, and professional opinions concerning the materials, fabrication, requirements, capabilities, and endurance of European and American Guns, for Naval, Sea-Coast, and Iron-Clad Warfare, and their Rifling, Projectiles, and Breech-Loading; also, results of experiments against armor, from official records, with an appendix referring to Gun-Cotton, Hooped Guns, etc., etc. 948 pages, 493 engravings, and 147 Tables of Results, etc. 8vo, half roan. 10 00

—— Railway Practice American and European Railway Practice in the economical Generation of Steam, including the Materials and Construction of Coal-burning Boilers, Combustion, the Variable Blast, Vaporization, Circulation, Superheating, Supplying and Heating Feed-water, etc., and the Adaptation of Wood and Coke-burning Engines to Coal-burning; and in Permanent Way, including Road-bed, Sleepers, Rails, Joint-fastenings, Street Railways, etc., etc. With 77 lithographed plates. Folio, cloth...................12 00

HOWARD (C. R.) Earthwork Mensuration on the Basis of the Prismoidal Formulæ. Containing simple and labor-saving method of obtaining Prismoidal Contents directly from End Areas. Illustrated by Examples, and accompanied by Plain Rules for Practical Uses. Illustrated. 8vo, cloth $1 50

INDUCTION-COILS. How Made and How Used. 63 illustrations. 16mo, boards 50

ISHERWOOD (B. F.) Engineering Precedents for Steam Machinery. Arranged in the most practical and useful manner for Engineers. With illustrations. Two volumes in one. 8vo, cloth... 2 50

JANNETTAZ (EDWARD). A Guide to the Determination of Rocks: being an Introduction to Lithology. Translated from the French by G. W. Plympton, Professor of Physical Science at Brooklyn Polytechnic Institute. 12mo, cloth.... 1 50

JEFFERS (Capt. W. N., U. S. N.) Nautical Surveying. Illustrated with 9 copperplates and 31 wood-cut illustrations. 8vo, cloth... 5 00

JONES (H. CHAPMAN). Text-Book of Experimental Organic Chemistry for Students. 18mo, cloth.............. 1 00

JOYNSON (F. H.) The Metals used in Construction: Iron, Steel, Bessemer Metal, etc., etc. Illustrated. 12mo, cloth. 75

—— Designing and Construction of Machine Gearing. Illustrated 8vo, cloth.. 2 00

KANSAS CITY BRIDGE (THE). With an account of the Regimen of the Missouri River, and a description of the methods used for Founding in that River. By O. Chanute, Chief-Engineer, and George Morrison, Assistant-Engineer. Illustrated with five lithographic views and twelve plates of plans. 4to, cloth.. 6 00

KING (W. H.) Lessons and Practical Notes on Steam, the Steam-Engine, Propellers, etc., etc , for young Marine Engineers, Students, and others. Revised by Chief-Engineer J. W. King, U. S. Navy. Nineteenth edition, enlarged. 8vo, cloth.. 2 00

KIRKWOOD (JAS. P.) Report on the Filtration of River Waters for the supply of Cities, as practised in Europe, made to the Board of Water Commissioners of the City of St. Louis. Illustrated by 30 double-plate engravings. 4to, cloth... 15 00

LARRABEE (C. S.) Cipher and Secret Letter and Telegraphic Code, with Hogg's Improvements. The most perfect secret code ever invented or discovered. Impossible to read without the key. 18mo, cloth............................... 1 00

LOCK (C. G.), WIGNER (G. W.), and HARLAND (R. H.) Sugar Growing and Refining. Treatise on the Culture of Sugar-Yielding Plants, and the Manufacture and Refining of Cane, Beet, and other sugars. 8vo, cloth, illustrated 12 00

LOCKWOOD (THOS. D.) Electricity, Magnetism, and Electro-Telegraphy. A Practical Guide for Students, Operators, and Inspectors. 8vo, cloth............................$2 50

LORING (A. E.) A Hand-Book on the Electro-Magnetic Telegraph. Paper boards...................... 50
Cloth 75
Morocco.. 1 00

MACCORD (Prof. C. W) A Practical Treatise on the Slide Valve by Eccentrics, examining by methods the action of the Eccentric upon the Slide-Valve, and explaining the practical processes of laying out the movements, adapting the valve for its various duties in the steam-engine. Second edition Illustrated. 4to, cloth 2 50

McCULLOCH (Prof. R S.) Elementary Treatise on the Mechanical Theory of Heat, and its application to Air and Steam Engines. 8vo, cloth......................... 3 50

MERRILL (Col. WM. E., U. S. A.) Iron Truss Bridges for Railroads. The method of calculating strains in Trusses, with a careful comparison of the most prominent Trusses, in reference to economy in combination, etc., etc. Illustrated. 4to, cloth 5 00

MICHAELIS (Capt. O. E., U. S. A.) The Le Boulenge Chronograph, with three lithograph folding plates of illustrations. 4to, cloth.................................. 3 00

MICHIE (Prof. P. S.) Elements of Wave Motion relating to Sound and Light. Text-Book for the U.S. Military Academy. 8vo, cloth, illustrated............................. 5 00

MINIFIE (WM.) Mechanical Drawing. A Text-Book of Geometrical Drawing for the use of Mechanics and Schools, in which the Definitions and Rules of Geometry are familiarly explained; the Practical Problems are arranged, from the most simple to the more complex, and in their description technicalities are avoided as much as possible. With illustrations for Drawing Plans, Sections, and Elevations of Railways and Machinery; an Introduction to Isometrical Drawing, and an Essay on Linear Perspective and Shadows. Illustrated with over 200 diagrams engraved on steel. Ninth edition. With an Appendix on the Theory and Application of Colors. 8vo, cloth 4 00

"It is the best work on Drawing that we have ever seen, and is especially a text-book of Geometrical Drawing for the use of Mechanics and Schools. No young Mechanic, such as a Machinist, Engineer, Cabinet-maker, Millwright, or Carpenter, should be without it."—*Scientific American*.

—— Geometrical Drawing. Abridged from the octavo edition, for the use of schools. Illustrated with forty-eight steel plates. Fifth edition. 12mo, cloth 2 00

MODERN METEOROLOGY. A Series of Six Lectures, delivered under the auspices of the Meteorological Society in 1878. Illustrated. 12mo, cloth.... $1 50

MORRIS (E.) Easy Rules for the Measurement of Earthworks, by Means of the Prismoidal Formula. 78 illustrations. 8vo, cloth... 1 50

MOTT (H. A , Jr.) A Practical Treatise on Chemistry (Qualitative and Quantitative Analysis), Stoichiometry, Blow-pipe Analysis, Mineralogy, Assaying, Pharmaceutical Preparations, Human Secretions, Specific Gravities, Weights and Measures, etc., etc., etc. New edition, 1883. 650 pages. 8vo, cloth... 4 00

NAQUET (A.) Legal Chemistry. A Guide to the Detection of Poisons, Falsification of Writings, Adulteration of Alimentary and Pharmaceutical Substances, Analysis of Ashes, and examination of Hair, Coins, Arms, and Stains, as applied to Chemical Jurisprudence, for the use of Chemists, Physicians, Lawyers, Pharmacists, and Experts. Translated, with additions, including a list of books and Memoirs on Toxicology, etc., from the French. By J. P. Battershall, Ph.D., with a preface by C. F. Chandler, Ph.D., M.D., LL.D. 12mo, cloth... 2 00

NOBLE (W. H.) Useful Tables. Pocket form, cloth... 50

NUGENT (E.) Treatise on Optics; or, Light and Sight, theoretically and practically treated, with the application to Fine Art and Industrial Pursuits. With 103 illustrations. 12mo, cloth... 1 50

PEIRCE (B.) System of Analytic Mechanics. 4to, cloth... 10 00

PLANE TABLE (THE). Its Uses in Topographical Surveying. From the Papers of the U. S. Coast Survey. Illustrated. cloth... 2 00

"This work gives a description of the Plane Table employed at the U. S. Coast Survey office, and the manner of using it."

PLATTNER. Manual of Qualitative and Quantitative Analysis with the Blow-Pipe. From the last German edition, revised and enlarged. By Prof. Th. Richter, of the Royal Saxon Mining Academy. Translated by Prof H. B. Cornwall, assisted by John H. Caswell. Illustrated with 87 woodcuts and one lithographic plate. Fourth edition, revised, 560 pages. 8vo, cloth... 5 00

PLYMPTON (Prof. GEO. W.) The Blow-Pipe. A Guide to its use in the Determination of Salts and Minerals. Compiled from various sources. 12mo, cloth... 1 50

—— The Aneroid Barometer: Its Construction and Use. Compiled from several sources. 16mo, boards, illustrated, 50
Morocco... 1 00

PLYMPTON (Prof. GEO. W.) The Star-Finder, or Planisphere, with Movable Horizon   Printed in colors on fine card-board, and in accordance with Proctor's Star Atlas... $1 00

POCKET LOGARITHMS, to Four Places of Decimals, including Logarithms of Numbers, and Logarithmic Sines and Tangents to Single Minutes. To which is added a Table of Natural Sines, Tangents, and Co-Tangents. 16mo, boards, 50
Morocco.. 1 00

POOK (S. M.) Method of Comparing the Lines and Draughting Vessels propelled by Sail or Steam. Including a chapter on Laying-off on the Mould-Loft Floor. 1 vol. 8vo, with illustrations, cloth... 5 00

POPE (F. L.) Modern Practice of the Electric Telegraph. A Hand-Book for Electricians and Operators. Eleventh edition, revised and enlarged, and fully illustrated. 8vo, cloth. 2 00

PRESCOTT (Prof. A. B) Outlines of Proximate Organic Analysis, for the Identification, Separation, and Quantitative Determination of the more commonly occurring Organic Compounds. 12mo, cloth.. 1 75

—— Chemical Examination of Alcoholic Liquors. A Manual of the Constituents of the Distilled Spirits and Fermented Liquors of Commerce, and their Qualitative and Quantitative Determinations. 12mo, cloth......................... 1 50

—— First Book in Qualitative Chemistry. Second edition. 12mo, cloth.. 1 50

PYNCHON (Prof. T. R.) Introduction to Chemical Physics, designed for the use of Academies, Colleges, and High-Schools. Illustrated with numerous engravings, and containing copious experiments with directions for preparing them. New edition, revised and enlarged, and illustrated by 269 illustrations on wood. Crown 8vo, cloth............. 3 00

RAMMELSBERG (C. F.) Guide to a Course of Quantitative Chemical Analysis, especially of Minerals and Furnace Products. Illustrated by Examples. Translated by J. Towler, M.D. 8vo, cloth.. 2 25

RANDALL (P. M.) Quartz Operator's Hand-Book. New edition, revised and enlarged, fully illustrated. 12mo, cloth... 2 00

RANKINE (W. J. M.) Applied Mechanics, comprising Principles of Statics, Cinematics, and Dynamics, and Theory of Structures, Mechanism, and Machines. Crown 8vo, cloth. Tenth edition. London................................. 5 00

—— A Manual of the Steam-Engine and other Prime Movers, with numerous tables and illustrations. Crown 8vo, cloth. Tenth edition. London, 1882............................... 5 00

—— A Selection from the Miscellaneous Scientific Papers of, with Memoir by P. G. Tait, and edited by W. J. Millar, C.E. 8vo, cloth. London, 1880.................................10 00

RANKINE (W. J. M.) A Manual of Machinery and Mill-work. Fourth edition. Crown 8vo. London, 1881 ....$5 00

——— Civil Engineering, comprising Engineering Surveys, Earthwork, Foundations, Masonry, Carpentry, Metal-works, Roads, Railways, Canals, Rivers, Water-works, Harbors, etc., with numerous tables and illustrations. Fourteenth edition, revised by E. F. Bamber, C.E. 8vo. London, 1883.... .... 6 50

——— Useful Rules and Tables for Architects, Builders, Carpenters, Coachbuilders, Engineers, Founders, Mechanics, Shipbuilders, Surveyors, Typefounders, Wheelwrights, etc. Sixth edition. Crown 8vo, cloth. London, 1883...... 4 00

——— and BAMBER (E. F.) A Mechanical Text-Book; or, Introduction to the Study of Mechanics and Engineering. 8vo, cloth. London, 1875 .... 3 50

RICE (Prof. J. M.) and JOHNSON (Prof. W. W.) On a New Method of Obtaining the Differentials of Functions, with especial reference to the Newtonian Conception of Rates or Velocities. 12mo, paper 50

ROGERS (Prof. H. D.) The Geology of Pennsylvania. A Government Survey, with a General View of the Geology of the United States, Essays on the Coal Formation and its Fossils, and a description of the Coal Fields of North America and Great Britain. Illustrated with Plates and Engravings in the text. 3 vols. 4to, cloth, with Portfolio of Maps. 30 00

ROEBLING (J. A.) Long and Short Span Railway Bridges. Illustrated with large copperplate engravings of plans and views. Imperial folio, cloth.. 25 00

ROSE (JOSHUA, M.E.) The Pattern-Maker's Assistant, embracing Lathe Work, Branch Work, Core Work, Sweep Work, and Practical Gear Constructions, the Preparation and Use of Tools, together with a large collection of useful and valuable Tables. Third edition. Illustrated with 250 engravings. 8vo, cloth 2 50

SABINE (ROBERT). History and Progress of the Electric Telegraph, with descriptions of some of the apparatus. Second edition, with additions, 12mo, cloth 1 25

SAELTZER (ALEX.) Treatise on Acoustics in connection with Ventilation. 12mo, cloth 1 00

SCHUMANN (F.) A Manual of Heating and Ventilation in its Practical Application for the use of Engineers and Architects, embracing a series of Tables and Formulæ for dimensions of heating, flow and return pipes for steam and hot-water boilers, flues, etc., etc. 12mo. Illustrated. Full roan 1 50

——— Formulas and Tables for Architects and Engineers in calculating the strains and capacity of structures in Iron and Wood. 12mo, morocco, tucks 2 50

D. VAN NOSTRAND'S PUBLICATIONS.

SAWYER (W. E.) Electric-Lighting by Incandescence, and its Application to Interior Illumination. A Practical Treatise. With 96 illustrations. Third edition. 8vo, cloth.$2 50

SCRIBNER (J. M.) Engineers' and Mechanics' Companion, comprising United States Weights and Measures, Mensuration of Superfices and Solids, Tables of Squares and Cubes, Square and Cube Roots, Circumference and Areas of Circles, the Mechanical Powers, Centres of Gravity, Gravitation of Bodies, Pendulums, Specific Gravity of Bodies, Strength, Weight, and Crush of Materials, Water-Wheels, Hydrostatics, Hydraulics, Statics, Centres of Percussion and Gyration, Friction Heat, Tables of the Weight of Metals, Scantling, etc., Steam and the Steam-Engine. Nineteenth edition, revised, 16mo, full morocco............ 1 50

—— Engineers', Contractors', and Surveyors' Pocket Table-Book. Comprising Logarithms of Numbers, Logarithmic Sines and Tangents, Natural Sines and Natural Tangents, the Traverse Table, and a full and complete set of Excavation and Embankment Tables, together with numerous other valuable tables for Engineers, etc. Eleventh edition, revised, 16mo, full morocco 1 50

SHELLEN (Dr. H.) Dynamo-Electric Machines. Translated, with much new matter on American practice, and many illustrations which now appear for the first time in print. 8vo, cloth, New York..........................(In press)

SHOCK (Chief-Eng. W. H.) Steam-Boilers: their Design, Construction, and Management. 450 pages text. Illustrated with 150 wood-cuts and 36 full-page plates (several double). Quarto. Illustrated. Half morocco......................15 00

SHUNK (W. F.) The Field Engineer. A handy book of practice in the Survey, Location, and Track-work of Railroads, containing a large collection of Rules and Tables, original and selected, applicable to both the Standard and Narrow Gauge, and prepared with special reference to the wants of the young Engineer. Third edition. 12mo, morocco, tucks... 2 50

SHIELDS (J. E.) Notes on Engineering Construction. Embracing Discussions of the Principles involved, and Descriptions of the Material employed in Tunnelling, Bridging, Canal and Road Building, etc., etc. 12mo, cloth 1 50

SHREVE (S. H.) A Treatise on the Strength of Bridges and Roofs. Comprising the determination of Algebraic formulas for strains in Horizontal, Inclined or Rafter, Triangular, Bowstring, Lenticular, and other Trusses, from fixed and moving loads, with practical applications and examples, for the use of Students and Engineers. 87 wood-cut illustrations. Third edition. 8vo, cloth............................. 3 50

www.ingramcontent.com/pod-product-compliance
Lightning Source LLC
Chambersburg PA
CBHW020831230426
43666CB00007B/1183